AF556869

Biochemical Methods
A Practical Approach

Biochemical Methods

A Practical Approach

Anusha Bhaskar

Alpha Science International Ltd.

Oxford, U.K.

Biochemical Methods: A Practical Approach
182 pgs.

Anusha Bhaskar
Associate Professor
Centre for Research and Development
PRIST University
Thanjavur

ALPHA SCIENCE INTERNATIONAL LTD.
7200 The Quorum, Oxford Business Park North
Garsington Road, Oxford OX4 2JZ, U.K.

www.alphasci.com

ISBN 978-1-84265-876-5

Printed in India

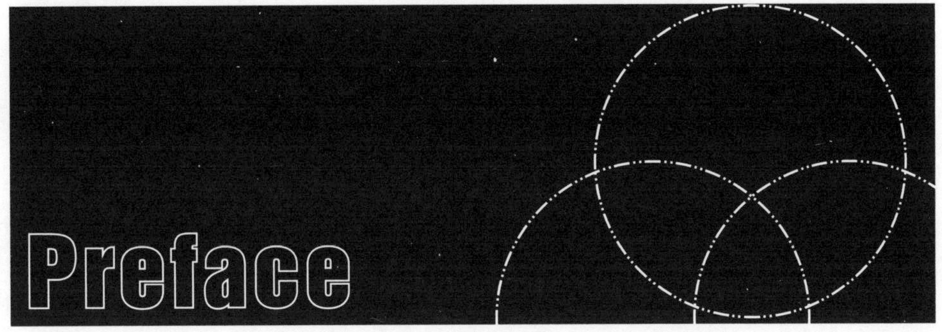

Preface

Biochemistry is the fundamental course to students of all disciplines of life sciences. This book has been written with a holistic approach, including various facets of practical biochemisty. It includes a fairly large number of practicals taking fully into account the meagre resources at the disposal of the instructors. This book will cater to the needs of graduate and post graduate students of various universities pursuing Biochemistry at the major level and the students of life sciences – Microbiology, Biotechnology, Food and Nutrition, Plant and Animal Sciences, Environmental Sciences, etc.

Each chapter in this book is organized giving importance to the principle of each experiment which is the fundamental knowledge required by the instructor and the students before performing an experiment. The techniques explained are those that can be performed in a laboratory that may not be very sophisticated. However, the ultimate aim of the book is to equip students with simple and easy protocols for analyses.

I hope that the manual will serve as a vehicle for the development of manipulative skills and techniques essential for understanding the complexity of biochemistry.

Anusha Bhaskar

Biochemistry is the fundamental course to students of all disciplines of life sciences. This book has been written with a holistic approach, including introductory/practical biochemistry. It includes a fairly large number of practicals suitable for laboratory experiments and the resources at the disposal of the institutes. This book will cater to the needs of graduate and post graduate students of various universities, studying Biochemistry at the graduate level and the students of life sciences, Microbiology, Biotechnology, Food and Nutrition, Plant and Animal Sciences, Environmental Sciences, etc.

Each chapter in this book is introduced by its importance to the principle of each experiment, which is the fundamental knowledge required by the instructor and the students before performing an experiment. The techniques explained are those that can be performed in a laboratory that may not be very sophisticated. However, the ultimate aim of the book is to acquaint students with simple and less common methods of analysis.

I hope that the manual will serve as a vehicle for the development of manipulative skills and techniques essential for understanding the concepts of biochemistry.

Anusha Bhaskar

Contents

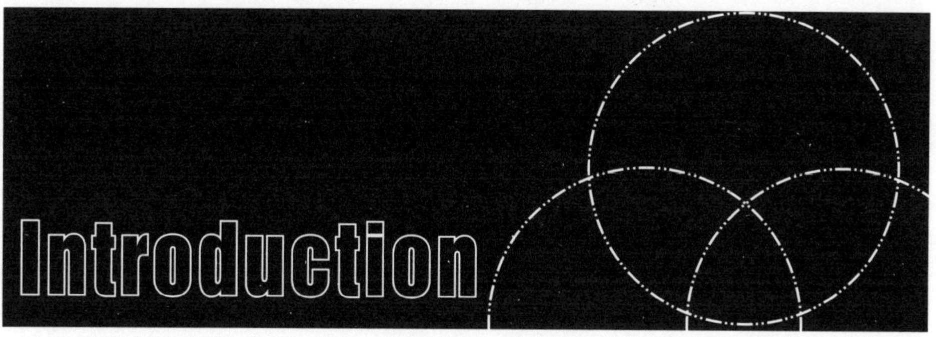

Introduction

Analysis is integral and indispensable to biochemical investigations. Methods of biochemical analyses need to be used whether one needs simple information such as protein concentrations, enzyme activity or for more complex studies like molecular biology.

Science has made enormous progress and developments in modern chemistry and biology have made biochemisty a fundamental science. As we move to an era of integrating science the students of life sciences require basic and formal training in biochemical laboratory techniques.

This book includes topics on qualitative and quantitative analyses of biomolecules, enzyme assays, food analyses, biochemical preparations, protocols to assay the extent of lipid peroxidation and antioxidants a current topic in biochemistry and fundamental experiments in molecular biology.

The protocols provided in the book have been established from years of class teaching. Majority of students in the undergraduate courses of biochemistry and other life sciences have difficulty in understanding and translating the instruction to protocols and standardizing it to the laboratory conditions. This may be a frustrating experience to the student. In order to make biochemistry laboratory more friendly and interesting, I have handpicked procedures which have been tested in our laboratory and provide a step-by-step protocol.

This book is intended to a laboratory guide not only to students and research scholars but also to the instructors of biochemistry laboratory.

Analysis is integral and indispensable to biochemical investigations. Methods of biochemical analyses need to be used whether one needs simple information such as protein concentrations, enzyme activity or for more complex studies like molecular biology.

Science has made enormous progress and developments in modern chemistry and biology have made biochemistry a fundamental science. As we move to an era of integrating science the students of life sciences require basic and formal training in biochemical laboratory techniques.

This book includes topics on qualitative and quantitative analyses of biomolecules, enzyme assays, food analyses, biochemical preparations, protocols to assay the extent of lipid peroxidation and antioxidants, a current topic in biochemistry and fundamental experiments in molecular biology.

The protocols provided in the book have been standardized from years of class teaching. Majority of students in the undergraduate courses of biochemistry and other life sciences have difficulty in understanding and translating the instructions of protocols and standardizing it to the laboratory conditions. This may be a frustrating experience to the student. In order to make biochemistry laboratory more friendly and interesting, I have highlighted procedures which have been standardized in our laboratory and provide a step by step protocol.

This book is intended for a laboratory guide not only to students and research scholars, but also to the instructors of biochemistry laboratory.

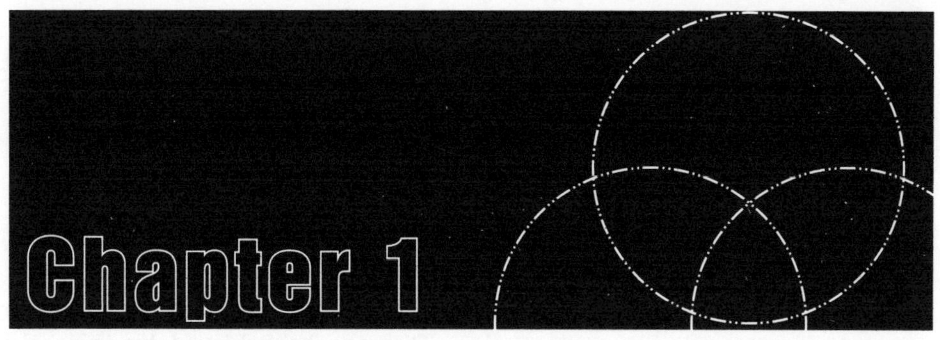

Chapter 1

Units and Measurements

Biochemistry employs the metric system of units that is based on the *Systéme International d' Unités* (SI units). To represent mass, volume and time, biochemists routinely use terms such as mole, milligram, microgram, parts per million, g%, liter, micro liter, minute, second, etc., to express the units in biochemical analysis.

Quantity	Name	Symbol
Length	Metre	m
Weight	Gram	g
Volume	Liter	L
Temperature	Celsius	°C
Amount of substance	Mole	mol
Solution	Molarity	M

1.1 REAGENTS SOLUTIONS

A solution is a soluble solid substance dissolved in a liquid. The former is called the solute and the latter the solvent. Reagent solutions are prepared in the laboratory for use in various laboratory procedures. These solutions are prepared according to the specifications provided in the test procedures.

(a) Weight / weight

g% = Weight of analyte in g per 100 g of substance.

mg% = Weight of analyte in mg per 100 g of substance.

(b) Weight / volume

g% (w/v) = Weight of solute in g per 100 ml of solvent

mg% (w/v) = Weight of solute in mg per 100 ml of solvent.

(c) Percent by volume

Biochemical analysis requires dilution of % solution (v/v) from higher concentration to a lower concentration. This can be achieved by using the following equation:

Example: Prepare 100 ml of 70% (v/v) ethanol from 95 % (v/v) ethanol.

95 (% of ethanol) × unknown volume (X) = 70 (% of ethanol) × Volume required (ml)

$$X = \frac{70 \times 100}{95} = 73.68 \text{ ml of } 95\% \text{ ethanol.}$$

Thus, to prepare 100 ml of 70% ethanol from 95% (v/v) ethanol – take 73. 68 ml of 95% (v/v) ethanol and dilute with 26.32 ml of distilled water.

Molar Solution (M)

1 mole (gram molecular weight) of solute in 1000 ml of solution is equivalent to 1 molar solution.

$$\text{Molarity} = \frac{\text{Weight of a solute in g/L of solution}}{\text{Mol. Wt of solute}}$$

Example: 58.5 g of NaCl in 1000 ml solution is equivalent to 1 M and 58.5 mg of NaCl in 1000 ml solution is equivalent to 1 mM.

1.2 MOLAL SOLUTION (M)

1 mole of solute dissolved in 1 kg of the solvent is equivalent to 1 molal solution.

$$\text{Molality} = \text{weight of a solute in g/kg of solvent}$$

Example:

1. How many grams of water must be used to dissolve 100 grams of Sucrose $C_{12}H_{22}O_{11}$ to prepare a .2 molal solution?
   (a) Determine moles of Sucrose in 100 grams.

   100 grams $C_{12}H_{22}O_{11}$ × 1 mole $C_{12}H_{22}O_{11}$/342 grams = 0.292 moles

   (b) Determine kilograms of solvent water from given molal and moles of solute molality = moles of Sucrose/kilograms of solvent

0.2 = 0.292/kg of water

kg of water = 0.292/.2 = 1.46 kg of water

(c) Convert kilograms into grams

1.46 kg × 1000 grams/1 kg = 1460 grams water

2. Determining the amount of solvent and solute for a given molality

   How would you prepare 5000 grams of a .5 molal NaOH solution?

   Determine grams of water in 1000 grams solution

   In 0.5 molal there is .5 mole NaOH in 1000 grams solvent

   0.5 mole NaOH × 40 grams/1 mole = 20 grams NaOH

   Total grams solution = grams of NaOH + grams of solvent

   1000 grams solution = 20 grams + grams solvent

   grams of solvent = 1000 – 20 = 980 grams solvent per 1000 grams solution

   5000 grams solution × 980 grams water/1000 grams solution = 4900 grams water

   grams NaOH = Total grams of solution - grams of solvent

   grams NaOH = 5000 – 4900 = 100 grams NaOH

1.3 NORMAL SOLUTION (N)

1 equivalent weight (in grams) of a substance in 1000 ml of solution is equivalent to 1 N.

Example: 40 g of sodium hydroxide (NaOH, 23 + 16 + 1 = 40) dissolved in water and made up to 1000 ml is equivalent to 1N.

1.4 IONIC STRENGTH

Ionic strength is a measure of the concentration of the total electrical charges existing in solution. It is equal to one half of the sum of the concentration (molarity) of each ion multiplied by square of its valency.

$$\mathbf{I} = \tfrac{1}{2}\,\Sigma M_i Z_i^2$$

Where

Σ = Sum of

M_i = Concentration of ion (Molarity)

Z_i = Net electrical charge of the ion (+ of –)

Example: The ionic strength of 150 mM solution of sodium chloride

$$I = ½ \Sigma M_i Z_i^2$$

150 mM NaCl = 0.15 M

Ionization of NaCl gives – Na^+ and Cl^-

Ionic strength of NaCl = ½ Σ[(0.15) × $(1)^2$ (for Na^+ ion)] + [(0.15) × $(1)^2$ (for Cl^- ion)] = **0.15**

1.5 DENSITY AND SPECIFIC GRAVITY

Density is defined as mass or weight per unit volume, usually given in terms of kg/l.

Specific gravity is a ratio of weight of a substance to the mass of an equal volume of water (density of water at 4°C is 1g/ml). Specific gravity is dimension less, thus numerically specific gravity is equal to density.

Example: Calculate the volume of conc. HCl (35% w/w*; density 1.18) required to prepare 1L of 0.5 M HCl.

This can be calculated by the following equation:

Weight of HCl required = Volume × density (g/ml) × fraction of HCl weight in 1 ml (weight %)

= 1000 ml × 1.18 × 0.35 = 413g

Number of moles of HCl = Weight/MW = 413/36.5 = 11.32

Thus, to prepare 1 L of 0.5 M HCl, the volume of Conc. HCl required is – 0.5 M/11.32 M = 0.0442 L, i.e. 44.2 ml.

* density of 1mL of water = 1 g by water

pH and Buffers

2.1 ACIDS AND BASES

The Brønsted–Lowry definition of acids and bases states that, an acid is a substance that can donate a hydrogen ion (proton) and base is a substance which can accept it.

The water molecule dissociates into H^+ and OH^-, it acts both as an acid and a base.

$$H^+ + \underset{\text{(base)}}{H_2O} \longrightarrow H_3O^+$$

$$\underset{\text{(acid)}}{H_2O} \longrightarrow H^+ + OH^-$$

pH

pH is mathematically defined as the negative log of hydrogen ion concentration (Eq. 2.1). In most cases, the activity of hydrogen ions in solution can be approximated by the Molar concentration of hydrogen ions ($[H^+]$) in a solution (Eq. 2.2).

Hence,

$$pH = -\log aH^+ \qquad (2.1)$$

becomes

$$pH \approx -\log [H^+] \qquad (2.2)$$

Using Eq. 2.2, it can be calculated that a pH of 4 corresponds to a hydrogen ion concentration of 10^{-4} Molar.

The benchmark for pH is actually the concentration of hydrogen ions in water at 25°C. At any temperature, water dissociates into hydrogen ions and hydroxide ions as shown in the following equilibrium reaction.

$$H_2O \leftrightarrow H^+ + OH^-$$

The rate of the equilibrium reaction above is described by an equilibrium constant K_W, which varies with temperature but is equal to 1.01×10^{-14} at 25°C. This constant is determined using Eq. 2.3.

$$K_W = [H^+][OH^-] \quad (2.3)$$

It is important to note that the value of K_W varies with temperature, and therefore so does the pH of a solution. For example, given the enthalpy of a reaction and the temperature at which the reaction occurs, the value of K_W for the reaction can be determined using the van't Hoff equation (Eq. 2.4).

$$\text{Ink}\ (T_2) = \text{Ink}\ (T_1) - (\Delta RH°/R)\ (T_2^{-1} - T_1^{-1}) \quad (2.4)$$

From calculated K_W values, the actual pH of water can be calculated at different temperatures. The pH of water at temperatures of 0°C, 25°C, and 100°C are shown in Table 2.1.

Table 2.1 pH of Water at Different Temperatures

Temperature	K_w	*pH*
0°C	0.114×10^{-14}	7.46
25°C	1.01×10^{-14}	7
100°C	49×10^{-14}	6.15

The pH of a solution measures the degree of acidity or alkalinity relative to the ionization of water sample. Pure water dissociates to yield 10^{-7} M of $[H^+]$ and $[OH^-]$ at 25°C; thus, the pH of water is neutral i.e. 7.

$$pH_{water} = -\log [H^+] = -\log 10^{-7} = 7$$

Most pH readings range from 0 to 14. Solutions with a higher $[H^+]$ than water (pH less than 7) are acidic; solutions with a lower $[H^+]$ than water (pH greater than 7) are basic or alkaline.

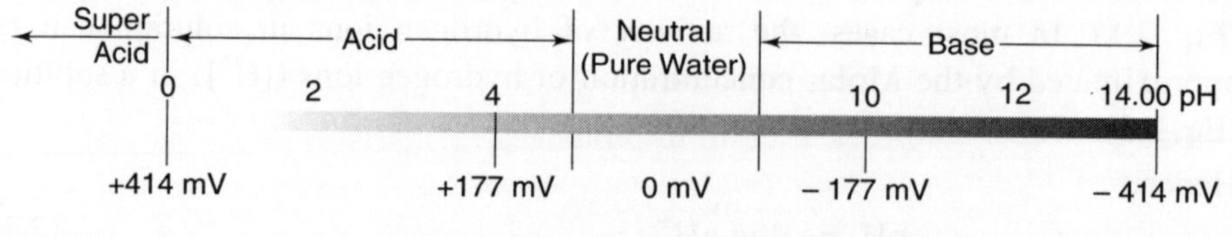

Figure 2.1

pH Measurement

A pH measurement is a determination of the activity of hydrogen ions in an aqueous solution. Many important properties of a solution can be determined from an accurate measurement of pH, including the acidity of a solution and the extent of a reaction in the solution. Many chemical processes and properties, such as the speed of a reaction and the solubility of a compound, can also depend greatly on the pH of a solution. In application ranging from industrial operations to biological processes, it is important to have an accurate and precise measurement of pH.

A solution's pH is commonly measured in a number of ways. Qualitative pH estimates are commonly performed using Litmus paper or an indicator in solution. Quantitative pH measurements are reliably taken using potentiometric electrodes. These electrodes simply monitor changes in voltage caused by changes in the activity of hydrogen ions in a solution. A simple potentiometric electrode of this type is shown in Figure 2.1.

Most modern pH electrodes consist of a single combination reference and sensing electrode instead of separate electrodes. This type of pH electrode is much easier to use and less expensive than the electrode pair in Figure 2.1. A combination electrode is functionally the same as an electrode pair.

Measuring pH involves comparing the potential of solutions with unknown $[H^+]$ to a known reference potential. pH meters convert the voltage ratio between a reference half-cell and a sensing half-cell to pH values.

In acidic or alkaline solutions, the voltage on the outer membrane surface changes proportionally to changes in $[H^+]$. The pH meter detects the change in potential and determines $[H^+]$ of the unknown by the Nernst equation:

$$E = E^o + (2.3RT)/nF \log \{\text{unknown } [H^+]/\text{internal } [H^+]\}$$

where

E = total potential difference (measured in mV);
E^o = reference potential;
R = gas constant;
T = temperature in Kelvin;
n = number of electrons;
F = Faraday's constant;
$[H^+]$ = hydrogen ion concentration.

This equation simplifies after all the constants are introduced.

$$E_{(pH)} = E_{(constant)} - 1.98158 \times 10^{-4} \text{ Volts/Kelvin} \times T \times pH$$

2.2 PREPARING BUFFERS AND BUFFER CAPACITY

A buffer solution is one that is resistant to change in pH when small amounts of strong acid or base are added. For example, when 0.01 mole of strong acid or base are added to distilled water, the pH drops to 2 with the acid and rises to 12 with the base. If the same amount of acid or base is added to an acetic acid – sodium acetate buffer, the pH may only change a fraction of a unit.

Buffers are important in many areas of biochemistry. When the pH must be controlled during the course of a reaction, the solutions are often buffered. This is the case in biochemistry when enzymes or proteins are being studied. Our blood is buffered to a pH of 7.4. Variations of a few tenths of a pH unit can cause illness or death. Acidosis is the condition when pH drops too low. Alkalosis results when the pH is higher than normal.

Two species are required in a buffer solution. One is capable of reacting with OH^- and the other will react with H_3O^+. The two species must not react with each other. Many buffers are prepared by combining a weak acid and its conjugate (acetic acid and sodium acetate) or a weak base and its conjugate (ammonia and ammonium chloride). In general, the pH range in which a buffer solution is effective is +/- one pH unit on either side of the pKa. The Henderson–Hasselbalch equation provides the information needed to prepare a buffer.

$$\text{pH} = \text{pKa} + \log \frac{[\text{conjugate base}]}{[\text{weak acid}]}$$

There is a limit to the amount of acid or base that can be added to a buffer solution before one of the components is used up. This limit is called the buffer capacity and is defined as the moles of acid or base necessary to change the pH of one liter of solution by one unit.

$$\text{Buffer Capacity} = \frac{(\text{number of moles of } OH^- \text{ or } H_3O^+ \text{ added})}{(\text{pH change})\ (\text{volume of buffer in L})}$$

In this experiment, thc Henderson-Hasselbalch equation will be used to determine the amount of acetic acid and sodium acetate required to prepare a series of buffer solutions. Once the buffer solutions have been prepared, their buffer capacity will be determined.

Basic Laws of Light Absorption – Beer-Lambert's Law

For a uniform absorbing medium (solution: solvent and solute molecules that absorb light) the proportion of light radiation passing through it is called the transmittance (*T*), and the proportion of light absorbed by molecules in the medium is absorbance (Abs).

Transmittance is defined as:

$T = I/I_o$ where;

I_o = intensity of the incident radiation entering the medium.

I = intensity of the transmitted radiation leaving the medium.

T is usually expressed as percent transmittance, % *T*:

$$\% T = I/I_o \times 100$$

The relationship between percent transmittance (% *T*) and absorbance (*A*) is given by the following equation:

$$A = 2\text{-log} (\% T)$$

On most spectrophotometers two scales are present, % *T* and A (absorbance) Absorbance has no units and varies from 0 to 2 (linear region for most substances is from 0.05 to 0.7). The Beer-Lambert's Law states that Abs is proportional to the concentration (c) of the absorbing molecules, the length of light-path through the medium and the molar extinction coefficient:

$A = \varepsilon cl$ where: ε = molar extinction coefficient for the absorbing material at wavelength (in units of mol/cm)

c = concentration of the absorbing solution (molar)
l = light path in the absorbing material (l = 1 cm for our purposes)

The Beer-Lambert's law may not be applicable to all solutions since solutions can ionize/polymerize at higher concentrations, or precipitate to give a turbid suspension that may increase or decrease the apparent absorbance. Further, the Beer-Lambert's law is most accurate between Abs of 0.05 to 0.70. Above 0.70, the measured Abs tends to underestimate the real Abs. Below 0.02 Abs many instruments are not accurate (remember that Abs is the log of a ratio). For example, a protein solution has an Abs of 1.42, but when it is diluted 1:4 it has an Abs of 0.45 meaning that the original, undiluted solution should have read Abs of $4 \times 0.45 = 1.80$ Abs.

3.1 UV/VISIBLE SPECTROSCOPY

The absorption spectrum (plural, spectra), or more correctly the absolute absorption spectrum, of a compound may be shown as a plot of the light absorbed by that compound against wavelength. Such a plot for a colored compound will have one or more absorption maxima ($\lambda_{max's}$) in the visible region of the spectrum (400 to 700 nm).

Absorption spectra in the ultraviolet (200 to 400 nm) and visible regions are due to energy transitions of both bonding and nonbonding outer electrons of the molecule. Usually delocalized electrons are involved such as the bonding electrons of C=C and the lone pairs of nitrogen and oxygen. Since most of the electrons in a molecule are in the ground state at room temperature, spectra in this region give information about this state and the next higher one. As the wavelengths of light absorbed are determined by the actual transitions occurring, specific absorption peaks may be recorded and related to known molecular substructures. The term chromophore is given to that part of a molecule that gives rise independently to distinct parts of an absorption spectrum, for example the carbonyl group. Conjugated double bonds lowers the energy required for electronic transitions and results in an increase in the wavelength at which a chromophore absorbs. This is referred to as a bathochromic shift, whereas a decrease in conjugation, caused for example by protonating a ring nitrogen atom, causes a hypochromic shift which leads to a decrease in wavelength. Hyperchromic and hypochromic effects refer to an increase and a decrease in absorbance respectively.

3.2 INSTRUMENTATION

To obtain an absorption spectrum, the absorbance of a substance must be measured at a series of wavelengths. Absorption in the visible and ultraviolet regions can be measured by a UV/visible spectrophotometer.

UV/Vis spectrometers consist of three basic components, (i) a light source and a mechanism to select a specific wavelength of light in the UV/visible region of the spectrum, (ii) a chamber where a cuvette containing a test solution can be introduced into the light path, and (iii) a photocell that can determine the amount of light absorbed by the sample (or the intensity of light transmitted through the sample).

The light source is usually a tungsten lamp for the visible region of the spectrum, and either a hydrogen or deuterium lamp for ultraviolet wavelengths.

Cuvettes are optically transparent cells that hold the material(s) under study and are used to introduce samples into the light path. A reference cuvette optically identical to, and containing the same solvent (and impurities) as the test cuvette is always required for setting the spectrophotometer to read zero absorbance at each wavelength used. For accurate work, the optical matching of the two cuvettes should always be checked. Glass and plastic absorb strongly below 310 nm and are not useful for measuring absorbance below that wavelength. Quartz or silica cells are used when measuring absorption of ultraviolet wavelengths by a solution since they are transparent to wavelengths greater than 180 nm.

When a cuvette is positioned in the light path it becomes an integral part of the instrument's optical system, so it should be treated with the same care given to other optical components. The use of scratched or contaminated cuvettes should be avoided since they reflect and/or absorb radiation that will give you inaccurate measurements. Also, bubbles, turbidity, fingerprints, or condensation on, or inside, cuvettes should be avoided since they will diminish the accuracy of readings.

3.3 COLORIMETRY

Many substances that do not absorb or have a very low absorbance in the visible region of the spectrum will react quantitatively with specific reagents to give a colored product that can be measured in the spectrophotometer. The absorbance (A) of such solutions can be plotted against the quantity or concentration (c) of the test substance producing the color. This graph is known as the calibration curve (or standard curve). Since it is often possible to produce comparatively high absorbances with relatively small amounts of material, colorimetry is widely used in biochemistry to assay a wide range of biologically important molecules.

Some important points to bear in mind when carrying out colorimetry are as follows:

(i) unlike straightforward spectrophotometry, colorimetry is a destructive technique; i.e., once reacted, the sample cannot be recovered;

(ii) a chromophore reflects the complementary color(s) that it absorbs, i.e. a yellow compound appears yellow because it absorbs blue light and therefore it must be estimated in the blue region of the spectrum;

(iii) colorimetric assays are usually most sensitive at the λ_{max} of the chromophore produced. If this is not known, the spectrum needs to be determined before commencing the assay;

(iv) reference cuvettes (i.e., blanks) should contain everything except the substance being assayed i.e., all the reagents in the same concentrations as in the test cuvettes;

(v) the reference cuvette and its contents require, if anything, more care because any error in these will be reflected in all of the values obtained;

(vi) assays should normally be performed in duplicate or triplicate, and individual values, not means, should be plotted. This procedure allows the experimenter to justifiably omit erroneous values from the calibration curve;

(vii) the "line of best fit" should be drawn through the data points, not necessarily the line that passes through the origin and the other points.

(viii) calibration curves may vary as batches of reagents and standards vary. Therefore, new calibration curves should be prepared each time an assay is run;

(ix) calibration curves should never be extrapolated beyond the highest absorbance value measured. It is always more accurate to repeat an assay at a concentration which falls within the most accurate region of the calibration curve (i.e., you may need to dilute a solution to accomplish this). This is usually from 0.05 to 0.70

3.4 VERIFICATION OF BEER'S LAW

Prepare a 2 mg/ml solution of potassium dichromate, take a series of clean test tubes and pipette out 0.5 – 2.5 ml of this solution and make up the volume to 5 ml using distilled water. Mix the solutions well. Measure the absorbance and transmittance of each tube using a blank that does not contain potassium dichromate using a blue filter (440 nm).

Draw two graphs taking the concentrations of dichromate (mg/ml) on X-axis versus absorbance or transmittance on the Y-axis. A straight line will indicate proportional increase in absorbance at 440 nm with increasing concentrations of potassium dichromate. Such a graph relating concentration to absorbance is called *standard graph*. From this graph, the concentration of an unknown solution can be determined if its absorbance or transmittance is known.

3.5 ABSORPTION SPECTRA

To get the absorption spectra of a compound prepare the solution of the compound (appropriate concentration). Turn on the instrument and set zero with the blank. Keep the test solution in a cuvette and measure the absorbance. Repeat the experiment by carrying the wavelength by 5 nm each time, adjust the zero with blank and measure the absorbance. Plot a graph with wavelength on X-axis and absorbance on Y-axis. This is called **absorption spectrum**. The peak maximum at any wavelength is the absorption maximum of the compound studied. Some compounds can have more than one absorption maxima.

3.6 MOLAR EXTINCTION COEFFICIENT

The extinction coefficient (ε) in the Beer-Lambert's equation is a measure of how strongly a compound in solution absorbs light at a particular wavelength.

Prepare a solution whose concentration is accurately known. Measure its absorbance at its absorption peak. Calculate the molar extinction coefficient by using the equation

$$\varepsilon = A/cl$$

The molar extinction coefficient is a constant for a given compound at the specified wavelength, and is maximum for a pure compound. Any impurity will decrease the value as measured. Extinction coefficients are useful numbers that allow researchers to calculate the concentration of a compound in solution (quantitative analysis).

By the same reasoning, if you know A, ε and l, we will be able to calculate the molar concentration (c) of a compound in solution.

(a) Calculate the molar concentration (c) of undiluted standard compound._

(b) Calculate the molar concentration of the diluted standard.

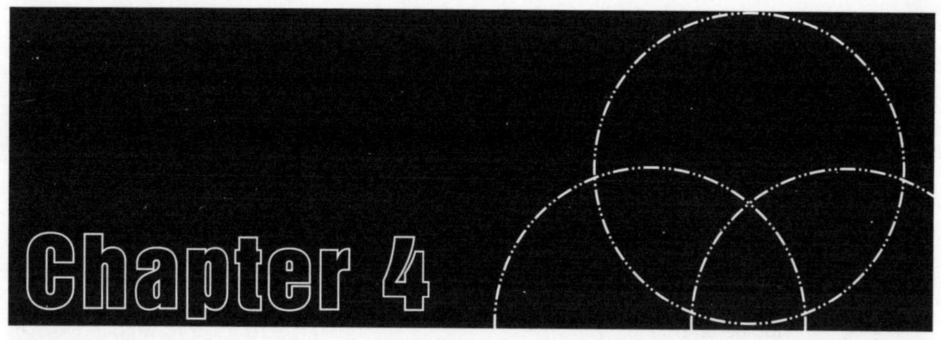

Carbohydrates

4.1 QUALITATIVE TESTS FOR CARBOHYDRATES

Qualitative tests are based on the following:

(a) test based on the reducing property of the aldehyde,

(b) those which make use of the dehydration of hydroxyl groups by strong acid and treating the resultant furfural with various compounds to form coloured products, and

(c) preparation of derivatives.

It is required to perform the test on a rationale as shown schematically.

MOLICH'S TEST

This is a test which confirms the presence of carbohydrates in the sample. Conc. H_2SO_4 hydrolyses glycosidic bonds to yield monosaccharides which in the presence of an acid get dehydrated to form furfural and its derivatives. These products react with sulphonate α-napthol to give a purple complex.

Reagents

1. α-napthol: 5% w/v in ethanol
2. Conc. sulphuric acid

Procedure

Take 2 ml of the test solution in a tube and add α-napthol solution. To this solution slowly add conc. sulphuric acid along the sides of the tube, two distinct layers will be formed and a violet coloured ring will be formed at the junction of the two layers. This indicates the presence of sugars.

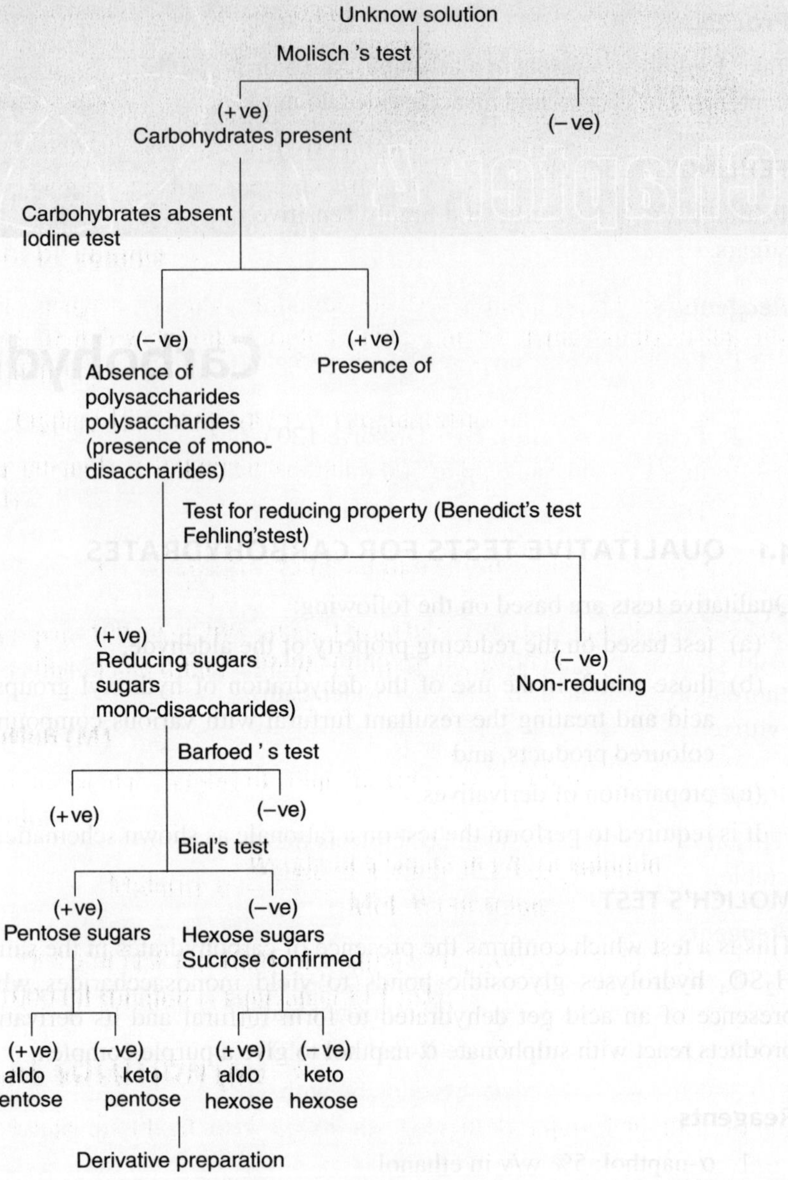

IODINE TEST

Iodine forms a complex with polysaccharides. Starch gives a blue colour while glycogen gives a reddish brown complex.

Reagents

(a) **Iodine solution:** Dissove 3g of potassium iodide in distilled water and add a few crystals of iodine until the solution becomes deep yellow in colour.

Procedure

Take 1 ml of the sample in a test tube. Add 5 drops of iodine solution and mix the contents. The mono- and disaccharides do not form any colour.

FEHLING'S TEST

Fehlings test is a specific and highly sensitive test for the detection of reducing sugars.

Reagents

1. **Fehling's solution A:** Dissolve 35 g of $CuSO_4 \cdot 5H_2O$ in water and make the volume to 500 ml.
2. **Fehling's solution B:** Dissolve 120 g of KOH and 173 g Na-K tartarate in water and make up to 500 ml.
3. **Fehling's reagent:** Mix equal volumes of solutions A and B just before use.

Procedure

Add a few drops of Fehling's reagent to 2 ml of the test solution. Mix the contents thoroughly and place in a boiling water bath. A red precipitate will be formed with reducing sugars.

BENEDICT'S TEST

This also works on the same principle as the above test, but the reaction is more stable.

Reagents

1. Benedict's reagent: Dissolve 173 g of sodium citrate and 100 g of anhydrous sodium carbonate in 600 ml of hot water. Dilute to 800 ml with water.
2. Dissolve 173 g of $CuSO_4 \cdot 5H_2O$ in 100 ml hot water.
3. Mix both the reagents slowly with constant stirring and make up to 1 litre.

Procedure

Add 0.5 – 1 ml of the test solution to 2 ml of Benedict's reagent. Shake the tubes thoroughly and place in a boiling water bath. Observe the appearance of a red precipitate. The precipitate is formed by reducing sugars.

PICRIC ACID TEST

Reducing sugars react with picric acid to form red coloured picramic acid.

Reagents

1. Saturated picric acid: Dissolve 13 g of picric acid in distilled water.
2. 10% Na_2CO_3 solution

Procedure

To 1 ml of sample add 1 ml of picric acid and heat in a boiling water bath. The appearance of a red coloured precipitate indicates the presence of reducing sugars.

BARFOED'S TEST

This test is specific for monosaccharides. Therefore, it is useful to distinguish monosaccharides from reducing disaccharides.

Reagents

Dissolve 13.3 g of copper acetate in 200 ml of water and add 1.8 ml of glacial acetic acid.

Procedure

Take 1 ml of the test solution and add 2 ml of Barfoed's reagent. Mix thoroughly and heat in a boiling water bath. Formation of a red precipitate indicates the presence of monosaccharides.

BIAL'S TEST

This test is used to determine pentoses. Furfural is formed in acid medium in the presence of orcinol and ferric ions it condenses to give a blue-green coloured complex

Reagents

Dissolve 1.5 g of orcinol in 100 ml of conc. HCl and add 2 ml of 10% ferric chloride solution to it.

Procedure

Add 2 ml of Bial's reagent to 2 ml of the test solution. Heat in a boiling water bath. Appearance of a blue – green coloured complex indicates the presence of pentoses.

SELIWANOFF'S TEST

Aldo- and keto sugars can be distinguished using this test. Ketoses undergo dehydration to give a furfural derivative which then condenses with resorcinol to form a red complex.

Reagents

0.05% of resorcinol in 3 N HCl.

Procedure

To 1 ml of the test solution add 2 ml of Seliwanoff's reagent and heat in a boiling water bath for 1 min. Appearance of deep cherry red colour indicates the presence of keto sugars.

MUCIC ACID TEST

This test is specific for galactose.

Reagents

Solid galactose

Solid glucose

Conc. HNO_3

Procedure

To 2 ml of the sample add 1 ml of conc. HNO_3. Heat the tube in boiling water bath for 1 h. Add 5 ml water and keep over night. Insoluble mucic acid will be formed if galactose is present.

PHYENYLHYDRAZINE TEST

The sugars with phenylhydrazine and sodium acetate give characteristic needle shaped crystals, unique for each sugar.

Reagents

Phenylhydrazine

10% sodium acetate

Procedure

To 1 ml of the test solution add a few drops of phenylhydrazine and sodium acetate. Mix well and heat in a boiling water bath. The time at which the sugars form derivatives with phenylhydrazine varies, for instance glucose and fructose – 5 min, lactose – 20 min, maltose – 10 min, galactose – 6-8 min. The yellow precipitate formed is viewed under a microscope. If you see:

Needle shaped crystals - glucose and fructose.

Ball shaped crystals – lactose

Sunflower shaped – maltose

Broken glass piece shaped – galactose

4.2 ESTIMATION OF GLUCOSE – BENEDICT'S TITRIMETRIC METHOD

Principle

Presence of reducing sugars form a red precipitate of cuprous oxide and sodium citrate in the reagent functions as a chelating agent.

$$Na_2CO_3 + 2H_2O \longrightarrow 2NaOH + H_2CO_3$$

$$2NaOH + CuSO_4 \longrightarrow Cu(OH)_2 + Na_2SO_4$$

$$Cu(OH)_2 \longrightarrow CuO + H_2O$$

$$\text{D-Glucose} + 2\,CuO \longrightarrow \text{D-gluconic acid} + \underset{\text{(Red ppt)}}{Cu_2O}$$

Reagents

Benedicts reagent: Weigh 200 g of sodium citrate, 100 g of sodium carbonate and 125 g of potassium thiocyanate and dissolve in 800 ml of distilled water. Filter the solution. Dissolve 15 g of copper sulphate separately in 100 ml of distilled and pour this solution into the former with constant stirring. Then add 5 ml of 5% potassium ferrocyanide solution to it and make upto 1 litre.

Standard glucose solution: Weigh 100 mg of glucose and dissolve in 100 ml of distilled water in a volumetric flask. From this 10 ml of the solution is made upto to 100 ml and this serves as the working standard.

Procedure

Pipette out 5 ml of Benedicts reagent into a clean conical flask. To this add 1 g of anhydrous sodium carbonate. Heat the contents to boiling and titrate against the glucose taken in the burette. The end point is the disappearance of blue colour and the appearance of a white precipitate of copper thiocyanate. Benedicts solution is standardized from the volume of glucose solution required to reduce 5 ml Benedicts reagent. Now the unknown sugar solution is taken in the burette and the titrations are repeated. The quantity of glucose in unknown solution is calculated.

Reference

Benedict, S.R.J. Amer. Med. Asso. 57: 1194, (1911).

4.3 ESTIMATION OF TOTAL SOLUBLE SUGARS BY FERRICYANIDE METHOD

Sugar is oxidized with alkaline ferricyanide and the ferrocyanide produced is measured photometrically after the conversion.

Reagents

1. Ferricyanide solution: Dissolve 0.5 g potassium ferricyanide in 1 litre of distilled water. Store in brown bottle.
2. Carbonate – cyanide reagent: Dissolve 5.3 g sodium carbonate and 0.65 g potassium cyanide in litre of distilled water.
3. Ferric ammonium sulphate solution: 1.5 g ferric ammonium sulphate in 1 litre of 0.05 N sulphuric acid.
4. Tungstate solution: 10% sodium tungstate.
5. Standard glucose 90 mg of glucose in 100 ml of benzoic acid.
6. N/12 sulphuric acid.

Procedure

Add 0.02 ml of sample to 0.2 ml of N/12 sulphuric acid and make up to 1.0 ml with water and mix thoroughly. To precipitate the proteins in the sample add 0.02 ml of sodium tungstate and mix the solution well. Allow to stand for 10 min and centrifuge. Remove 1 ml of the supernatant, mix with 1 ml each of ferricyanide and cyanide – carbonate solution. Heat the solution in a boiling water bath for 15 min and cool immediately under running water. Add 5 ml of ferric ammonium sulphate (60 ml ferric ammonium sulphate and 0.12 ml Tween 20), mix the solution and measure the absorbance at 690 nm in a spectrophotometer after 10 min but within 30 min. The standard glucose is diluted as per the range required and treated as above.

Reference

Hagedorn, H.C. and Jensen,B.N. (1923), Biochem. Z. 125: 46.

4.4 ESTIMATION OF REDUCING SUGARS - NELSON AND SOMOGYI METHOD

Blood proteins are precipitated by zinc hydroxide. The filtrate is heated with alkaline copper reagent and the reduced copper formed is treated with arseno-molybdate reagent resulting in the formation of violet colour.

Reagents

1. 5 % Zinc sulphate solution.
2. 0.3 N Barium hydroxide

   These two solutions should be so adjusted that 5 ml zinc sulphate requires 4.7 to 4.8 ml barium hydroxide for complete neutralization.
3. Alkaline copper reagent.

   Solution: Dissolve 25 g anhydrous sodium carbonate, 25 g Rochelle salt, 20g sodium bicarbonate and 200 g anhydrous sodium sulphate in 800 ml water and dilute to 1 liter. Store the solution at room temperature and filter before use.

Solution: B: 15% $CuSO_4 .5H_2O$ containing 1–2 drops of sulphuric acid.

Mix solutions A and B in the ratio 25:1 before use.

4. Aresnomolybdate colour reagent: Ammonium molybdate 25 g is dissolved in 150 ml water, 21 ml concentrated sulphuric acid is added and stirred. It is then placed in an incubator at 37 °C for 24 – 48 h and stored in a glass stoppered brown bottle.
5. Standard glucose: 100 mg of glucose dissolved in 100 ml of benzoic acid.

Procedure

Add 0.1 ml of blood to 3.5 ml of water in a test tube and mix well. To this add 0.2 ml of 0.3 N barium hydroxide, the solution turns brown, now add 0.2 ml of zinc sulphate and mix well. Centrifuge the tubes after 10 – 15 min and filter through Whattman filter paper.

Transfer 1 ml aliquot of the filtrate into 2 separate tubes and add 1 ml of alkaline copper reagent. Cover the tubes with glass marbles and place them in a boiling water bath for 20 min. Cool the tubes under running water and add 1 ml of aresnomolybdate reagent and dilute this solution to 25 ml with distilled water. Process standards a reagent blank and the absorbance is read at 500 nm.

References

1. *Somogyi, M. (1945), J. Biol. Chem. 160: 61.*
2. *Somogyi, M. (1952), J. Biol. Chem. 195: 19.*

4.5 REDUCING SUGARS BY DINITROSALICYLATE (DNS) METHOD

Reducing sugars convert dinitrosalicylate under alkaline condition to amino-nitrosalicylate, an orange-yellowish compound that has an absorption at 540 nm.

Reagents

1. DNS reagent: Dissolve 1 g of DNS, 200 mg of phenol and 50 mg of sodium sulphite in 100 ml of 1% NaOH.
2. Potassium – sodium tartarate: Dissolve 40g in 100 ml of water.
3. Standard maltose solution: Weigh 200 mg of maltose and make upto 100 ml in a volumetric flask.

Procedure

To 1 ml of the sample and standard sugar in the range 0.2 – 2 mg of maltose add 2 ml of DNS reagent. Heat the tubes in a boiling water bath for 5 minutes, while the tubes are still warm, add 1 ml of 40% potassium-sodium tartarate solution. Cool the tubes at room temperature and add 7 ml of distilled water and mix well. Read the absorbance at 540 nm.

Reference

Miller, G.I. (1972), Anal. Chem. 31: 426.

4.6 ESTIMATION OF BLOOD GLUCOSE BY O-TOLUDINE METHOD

Glucose reacts with o-toludine in acetic acid to give a Schiff's base.

Reagents

1. Trichloroacetic acid (TCA) – 10% in water.
2. O-toludine reagent- Dissolve 2.4 g boric acid and 2.5 g thiourea in 100 ml of a solution containing water, glacial acetic acid and freshly redistilled o-toludine in the ratio 10:75:15 (v/v).
3. Standard glucose: Dissolve 100 mg of D-glucose in 100 ml of 0.1% benzoic acid. 10 ml of this solution is diluted to 10 ml to get a working standard containing 100 μg glucose/ml.

Procedure

Mix 0.1 ml of blood with 3.0 ml of 10% TCA to precipitate the proteins. To one ml of the supernatant add 4 ml of o-toludine reagent. Pipette out standard solutions of glucose (25 – 100 μg) and make up to 1 ml with distilled water, blank contained 1 ml distilled water. To all the tubes add 4 ml O-toludine reagent. Heat the tubes in a boiling water bath for 15 min and the blue colour developed is read at 640 nm.

Glucose values are expressed as mg/dl.

Reference

Sasaki, P. and Matsui, S. (1972), Rhinsho Kagaku, 1: 346.

4.7 GLUCOSE ESTIMATION BY GLUCOSE OXIDASE METHOD

The glucose oxidase reaction in conjunction with an auxiliary reaction has been widely used for the determination of glucose in biological fluids. Many different auxiliary reactions have been developed in order to improve the overall specificity of glucose oxidase. The method utilized in this reagent is based on the hydrogen peroxide indicator reaction, which couples 4-aminoantipyrine to a phenolic compound.

$$\text{Glucose} + O_2 + H_2O \xrightarrow{\text{Glucose oxidase}} \text{Gluconic acid} + H_2O_2$$

$$H_2O_2 + HBA + 4\text{-}AAP \xrightarrow{\text{Peroxidase}} \text{Quinoneimine dye} + H_2O$$

Glucose is oxidized to gluconic acid and hydrogen peroxide. The hydrogen peroxide reacts in the presence of peroxidase with HBA and 4-aminoantipyrine forming a red quinoneiminc dye. The intensity of the colour formed is proportional to the glucose concentration and can be measured colorimetrically between 460 and 560 nm.

Reagents

Glucose oxidase enzyme: 20,000 U/litre

Peroxidase: 250 U/litre

4-aminoantipyrine: 0.5 mmol/l

4-hydroxybenzoic acid: 10 mmol/l

Phosphate buffer: 110 mmol/l

The reagent is the mixture of all the components in the right proportions.

Glucose standard: Dissolve100 mg of glucose in 100 ml of distilled water. A working standard is prepared by diluting 10 ml of the stock to 100 ml with water.

Procedure

Pipette out standard glucose in the range (10-50μg) in test tubes and also take the test sample. Add 1.5 ml of the reagent to all the tubes. Add 0.01 ml of water to all the tubes including thc blank, which is prepared by taking the reagent but contains no glucose. The absorbance is read at 500 nm and the unknown glucose is estimated.

References

1. *Farrance, I. (1987), Clin. Biochem. Reviews 8: 55.*
2. *Trinder, P. (1969), Ann. Clin. Biochem 6:24.*

4.8 ESTIMATION OF TOTAL SUGARS BY PHENOL SULPHURIC ACID METHOD

Sugars undergo dehydration in the presence of sulphuric acid to furfural or hydroxymethyl furfural that condenses with phenol to form a yellowish – orange coloured compounds with an absorption maxima at 490 nm.

Reagents

1. Phenol reagent: 5% phenol in distilled water.

2. Standard stock sugar solution: Dissolve 100 mg of mannose or galactose or glucose in 100 ml of water. Working standard is prepared by diluting 10 ml to 100 ml with water.

Procedure

To 2 ml of standard solution (10 – 100 µg) add 1 ml of phenol reagent, followed by rapid addition of 5 ml of concentrated sulphuric acid. The sample is treated similarly and after 30 min read the yellowish – orange colour developed against a reagent blank at 490 nm.

Reference

Dubois, M., Gilles, K.A., Hamilton, J.K., Rebers, P.A. and Smith, F. (1956), Anal. Chem. 28: 350.

4.9 ESTIMATION OF STARCH IN PLANT TISSUE

Sample is treated repeatedly with hot 80% alcohol. The residue rich in starch is solubilized with perchloric acid. Filtered perchloric acid-extract is treated with anthrone - sulphuric acid to determine glucose.

Reagents

1. Anthrone – sulphuric acid reagent: 0.2% anthrone – Dissolve 200 mg in 100 ml of sulphuric acid.
2. 80% ethyl alcohol
3. 52% perchloric acid

Procedure

Extraction of sugars and starch: Transfer 0.2 g of sample into a 50 ml centrifuge tube. Add a few drops of 80% alcohol to wet the sample and prevent clumping, add 5 ml of water and vortex thoroughly. Add 25 ml of hot 80% ethanol, mix well and centrifuge after 5 min of standing. Decant and discard the alcoholic solution. Add 30 ml of fresh hot 80% ethanol, stir and centrifuge as before. Discard the alcoholic solution. Repeat this washing twice for a total of four washings or until a test with anthrone is negative.

To the residue after final centrifugation add 5 ml of water, cool in ice-cold water and while stirring add 6.5 ml of diluted perchloric acid reagent, stir for about 5 minutes with a glass rod, keep for 15 min stirring occasionally. Add 20 ml of water and centrifuge. Pour the aqueous starch solution into a 100 ml volumetric flask cooled in ice water and stir while adding 6.5 ml of diluted perchloric acid reagent. Solubilize as before for 30 min at 0°C with occasional stirring and wash the contents of the tube into a 100 ml flask containing the first extract. Dilute the combined solution to 100 ml and filter.

Dilute 5 to 10 ml of the filtered starch solution to 500 ml or to contain 25 – 100 µg of per starch per 5 ml of solution. Pipette 5 ml of the dilute solution into a test tube, cool in a water bath and add 10 ml freshly prepared anthrone reagent. After the anthrone reagent has been added to all the tubes, mix each one thoroughly and heat them together for 7 minutes. Read the colour in a photocolorimeter at 630 nm.

Prepare a standard curve every time, using 20 -100 µg glucose containing the same amount of perchloric acid as that in starch aliquot and use this calibrated curve to obtain the yield of glucose from starch. Multiply the glucose value by 0.9 to convert into starch value.

Reference

Scott, T.A. and Melvin, E.H. (1953), Anal. Chem. 25: 1656.

4.10 ESTIMATION OF LIVER GLYCOGEN

Glycogen is hydrolyzed to glucose and the glucose thus formed is estimated.

Reagents

1. 30% KOH
2. 95% and 60% ethanol
3. 2N sulphuric acid
4. 6N NaOH
5. Phenol red indicator

Procedure

Wash the liver tissue and transfer it to a weighed stoppered test tube containing 30% KOH and weigh again. Adjust the amount of alkali to get 2 ml per gram of liver and digest the tissue in a boiling water bath for 1 h and 30 min and cool in ice-cold water. Add 2 volumes of 95% ethanol and heat the mixture to boiling. Allow it to stand overnight in the cold, centrifuge and dissolve the precipitate in 5–10 ml of warm water. Reprecipitate the glycogen with 2 volumes of 95% ethanol. Centrifuge and wash the precipitate with 60 % ethanol.

Add 2 ml of 2N H_2SO_4 per gram of liver and hydrolyze in a boiling water bath for 3–4h. Neutralize the solution with sodium hydroxide using phenol red as indicator, make up to required volume and filter. Glucose is estimated in an aliquot using anthrone or Nelson Somogyi method. To convert glucose to glycogen it is multiplied by a factor of 0.93.

4.11 ESTIMATION OF BLOOD LACTIC ACID

In this method a protein free filtrate is treated with copper sulphate and solid calcium hydroxide to remove glucose. A portion of the filtrate from this process is treated with sulphuric acid to convert the lactic acid into acetaldehyde, the amount of which is measured by means of the purple colour given by p-hydroxydiphenyl in the presence of copper.

Reagents

1. Copper sulphate 20% solution
2. Copper sulphate 4% solution
3. Concentrated sulphuric acid
4. p-hydroxydiphenyl solution: Dissolve 1.5 g of p-hydroxydiphenyl in 100 ml of 0.5% solution of sodium hydroxide.
5. Standard solution of zinc or lithium lactate: Dissolve 213 mg of pure dry lithium lactate in 100 ml of water, add 1ml of conc. sulphuric acid and make upto 1 litre with water. Dilute 1 in 20 to obtain the working standard.

Procedure

To 0.2 ml of blood add 3 ml of 10% tricholoracetic acid to precipitate the proteins. Take 2 ml of the protein free filtrate. To this add 1 ml of 20% copper sulphate solution and make up to 10 ml with water. Add about 1g of powdered calcium hydroxide and shake vigorously. Stand at room temperature for at least half an hour, shaking occasionally, and then centrifuge. Pipette 1 ml of supernatant fluid into a test tube and add 0.05 ml of 4% copper sulphate solution, add 6 ml of conc. sulphuric acid and mix well. Then place the tube upright in a boiling water bath for five min remove the tube and cool and add exactly 0.1 ml of the p-hydroxydiphenyl. Disperse the precipitated reagent quickly through the solution and place the tube in a bath at 30°C for 30 min, redispersing the precipitated reagent at least once during this time. Place in the boiling water bath for ninety seconds, cool to room temperature, and read at 560 nm. Put up a reagent blank using water instead of protein-free filtrate, and a standard in the range of 10 – 50 µg.

Reference

Barker, S.B. and Summerson, W.H. (1941), J. Biol. Chem., 138: 535.

4.12 ESTIMATION OF PYRUVATE

Pyruvic acid gives a colour with dinitophenyl hydrazine (DNPH) in the presence of sodium hydroxide.

Reagents

1. Dinitrophenyl hydrazine reagent: 100 mg of DNPH is dissolved in 100 ml of 0.2 N HCl.
2. Sodium hydroxide solution: 1.5 N
3. Stock pyruvate solution: Dissolve 125 mg of sodium pyruvate in 10 ml of distilled water. Dilute 10 ml of the stock solution to get the working standard.

Procedure

Pipette out 10 – 50 μg of standard pyruvic acid solution and 0.1 ml of the unknown solution in different test tubes. Make up to 3 ml with distilled water. Add 1 ml of DNPH reagent to all the tubes followed by 5 ml of 1.5 N NaOH. The purple colour developed is read at 540 nm. From the standard graph the amount of unknown pyruvate is calculated.

Reference

Lu, G.D. (1939), Biochem. J., 33: 249.

4.13 PROTEIN BOUND CARBOHYDRATES (GLYCOPROTEINS)

Precipitation

The protein bound carbohydrate components are precipitated using phosphotungstic acid. To 2 ml of the sample extract add 5 ml of 1% phosphotungstic acid in 0.5 N HCl. The precipitate is washed with 3 ml of 5% TCA. The pellets obtained are dissolved in 2 ml of 0.1N sodium hydroxide.

This extract is used for the estimation of protein bound hexose, hexosamine and sialic acid.

Protein Bound Hexose

Reagents

1. 4N HCl
2. Orcinol – sulphuric acid reagent. Dissolve 1.6 g of orcinol in 100 ml of distilled water. Mix 1.0 volume of this with 7.5 volumes of sulphuric acid –water mixture (in the ratio 3:2 v/v) before use.
3. Standard solution: Dissolve 50 mg galactose and 50 mg of mannose in 100 ml distilled water. Dilute this solution 1:10 to give a concentration of hexose.

Procedure

Add 0.5 ml of the neutralized sample to 0.5ml of distilled water. Add 7.5 ml of orcinol - sulphuric acid reagent and keep the tubes in an ice bath, gently shake the tubes and heat for 15 min in a water bath at 80°C. Standard solutions containing 20-100μg of hexose and reagent blank containing 1 ml of distilled water are processed simultaneously. The tubes are cooled to room temperature and the colour developed is read at 540 nm.

Protein bound hexose levels are expressed as μg/ mg protein of the sample suspension.

Protein Bound Hexosamine

One milliliter of the extract is hydrolyzed with 2.5 ml of 3N HCl for 6 hours after which it is neutralized with 6N NaOH.

Reagents

1. Acetyl acetone reagent:

   Reagent A: 1M trisodium phosphate

   Reagent B: 0.5 N potassium tetraborate

   Prepare 3.5% acetyl acetone by mixing reagent A and B in the ratio 98:2 (v/v).
2. Erhlich reagent: Dissolve 320 mg of p-dimethyl amino benzaldehyde in 24 ml of isopropanol – Conc HCl mixture in the ratio 7:1 (v/v).
3. Standard Galactosmine: Dissolve 10 mg of galactosmaine hydrochloride in 100 ml of distilled water. This contains 100 μg/ml.

Procedure

Add 0.6 ml of acetyl acetone reagent to 0.8 ml of neutralized sample. Heat the tubes in a boiling water bath for 30 min. Add 0.2 ml of Erlich's reagent after cooling and vortex. The absorbance is read at 540 nm. Standard galatosamine in the range 20–100 μg in 0.8 ml total volume and blank containing 0.8 ml distilled water were processed simultaneously. The amount of hexosamine present is expressed as μg/mg protein.

Protein Bound Sialic Acid

An aliquot of the extract is hydrolysed with 0.1N sulphuric acid at 80°C for 60 min and the sialic acid in the hydrolysate is estimated.

Reagents

1. Periodic acid reagent: Dissolve 14 g periodic acid in 100 ml of 0.1N sulphuric acid.

2. 4% sodium metaarsenite in 100 ml of 0.5N HCl.
3. TBA (0.1N): Dissolve 144 mg of TBA in 10 ml of distilled water. Adjust the pH of the solution to 9.0 with 6N NaOH and use the reagent when fresh.
4. Acidified butanol: Mix 95 ml butanol with 50 ml of HCl.
4. Standard solution: Dissolve 10 mg of N-acetyl neuraminic acid in 100 ml water.

Procedure

To 1 ml of the neutralized sample add 0.25 ml of periodic acid reagent and incubate at 37°C for 30 min. Vortex the tubes well, add 2 ml of TBA and heat in a boiling water bath for 8 min. Cool the tubes and add 5 ml of acidified butanol. Shake the tubes vigorously until the solution seperates into the aqueous and the butanolic phase. Separate the pink butanolic phase and read its absorbance at 540 nm against a reagent blank. Process standard solutions containing 10-50 μg of N-acetyl neruaminic acid similarly. Sialic acid is expressed as μg/mg protein.

References

1. *Tettamanti, G., Bonali, F., Marchesini, S. and Zambothi, V. (1973), Biochim. Biophys. Acta. 296: 160.*
2. *Niebes, P. (1972), Clin. Chim. Acta. 42: 399.*
3. *Wagner, W.D. (1979), Anal. Biochem. 94: 394.*

4.14 PAPER CHROMATOGRAPHIC SEPARATION OF CARBOHYDRATES

Principle

Sugars in a given mixture or sample are separated on the basis of difference in their solubilities and hence difference partition coefficients in a binary solvent system. The sugars with higher solubilities in stationary phase move slowly as compared to those with higher solubilities in mobile phase. The sugars thus separated are detected by spraying the colour reagents.

Solvents

Water – saturated phenol and 1% ammonia

n-butanol – acetic acid – water (4:1:5 v/v)

isopropanol – pyridine – water – acetic acid (8:8:4:1v/v)

The solvents are mixed in the right proportion in a separating funnel and shaken well. The solution appears to be separated into two phases, the aqueous phase at the bottom and the organic phase on the top.

Spray Reagents

Ammoniacal silver nitrate: Add equal volumes of NH_4OH to saturated solution $AgNO_3$ and dilute with methanol to give a final concentration of 0.3 M. After spraying the developed chromatograms place them in an oven for 5-10 min, when the reducing sugars appear as brown spots.

Alkaline permanganate: Prepare a 1% solution of $KMnO_4$ containing 2% Na_2CO_3. After spraying with this mixture, the chromatograms are kept at 100°C for a few minutes, when the sugar spots appear as yellow spots in purple background.

Aniline diphenylamine reagent: Mix 5 volumes of 1% aniline and 5 volumes of 1% diphenylamine in acetone with 1 volume of 85% phosphoric acid. After spraying the dried chromatograms with this solution the spots are visualized by heating the paper at 100°C for a few minutes.

Resorcinol reagent: Mix 1% ethanolic solution of resorcinol and 0.2 N HCl (1:1 v/v). Spray the dried chromatograms and visualize the spots by heating at 90°C.

Procedure

Take Whatman No. 1 filter paper sheet of appropriate size. Draw a line at the base 2 cm from the bottom breadthwise. And mark spots on the base line with a distance of 2.5 cm between the adjacent spots. Apply 2 µl of the sample using a fine capillary tube. The diameter of the spots should be kept as small as possible. Allow the spots to dry, now roll the paper into a cylinder, such that the spots face the outer side. Place the paper in the cylinder containing the running aqueous phase of the solvent system. This chamber should be previously saturated with the organic phase. The saturation should be complete in order to have an effective run.

Allow the solvent to rise to three fourths of the paper. Remove it and air dry. Spray the paper with the detecting solution it may sometimes need warming in an oven for a few minutes for the spots to develop. The sugars separate into clear zones.

The sugars are identified on the basis of the Rf values.

$$Rf = \frac{\text{Distance moved by the solute}}{\text{Distance moved by the solvent front}}$$

Reference

Partridge, S.M., and Westall, R.G. (1948), Biochem. J., 42: 238.

Qualitative Tests for Amino Acids

5.1 AMINO ACIDS AND PROTEINS

Qualitative tests are very useful in detecting the presence of amino acids and these are largely dependent on the nature of the groups present in the amino acids. The test are not used to greater extent in the present day, however, they may give us an idea about the different amino acids present in a protein sample.

5.2 ESTIMATION OF AMINO ACIDS AND PROTEINS

Ninhydrin Test

This is a general test to show the presence of amino acids in the sample. The test is based on a reaction between α-amino group of free amino acid and ninhydrin. Ninhydrin is a powerful oxidizing agent and in its presence, amino acids undergo oxidative deamination liberating ammonia, carbon dioxide, a corresponding aldehyde and reduced form of ninhydrin. The ammonia formed reacts with ninhydrin and its reduced product to give a blue substance diketohydrin (Ruhemann's purple).

Reagents

Ninhydrin reagent: 0.2% solution in acetone.

Procedure

Take 1 ml of the test solution and add 2-5 drops of ninhydrin reagent. Mix well and keep in a boiling water bath for 5 min. The amino acids give a purple or violet-blue colour. While imino acids (like praline) give a yellow colour.

Xanthoproteic Test

Aromatic amino acids (tyrosine, tryptophan and phenylalanine) form a yellow nitro derivative on heating with conc. HNO_3 and the salt of the derivative is orange in colour.

Reagents

Conc. HNO_3

Sodium hydroxide solution: 40%

Procedure

Add 1 ml of conc. HNO_3 to 1 ml of test solution. Cool and slowly add NaOH till the solution becomes alkaline. Appearance of orange red colour denotes the presence of aromatic amino acids.

Millon's Test

Tyrosine reacts with Millon's reagent.

Reagents

Millon's reagent: Dissolve 15 g of mercuric sulphate in 15% H_2SO_4.

Sodium nitrite: 1% solution

Procedure

Add 2- 5 drops of Millon's reagent into 1 ml of test solution. Mix well and place the tube in a boiling water bath for 10 min. Cool and add 1 ml of $NaNO_2$ solution. Appearance of a red colour indicates the presence of tyrosine.

Hopkin's –Cole Test

The indole groups of tryptophan reacts with glyoxylic acid in the presence of conc. H_2SO_4 to give a purple colour. Glyoxlyic acid is prepared by reducing oxalic acid with magnesium powder. Glacial acetic acid exposed to sunlight can be used as glyoxylic acid.

Reagents

Glyoxylic acid: Take the required amount of glacial acetic acid and leave it exposed to sunlight for a few minutes.

Procedure

To the test solution add 1 ml of the reagent. Mix well and add conc. H_2SO_4 slowly along the sides of the tube. A violet ring at the interface of the two liquids shows the presence of tryptophan.

Pauly's Test

Diazotized sulphanilic acid couples with amines, phenols and imidazoles present in the amino acids.

Reagents

Sulphanilic acid: 1% sulphanillic acid in 10% HCl.

Sodium nitrite: 5% solution.

Sodium carbonate: 1% solution

Procedure

Mix 1ml of sulphanilic acid (1% solution in 10% HCl) with 2 ml of test solution and cool on ice. Add 1 ml of 5% sodium nitrite solution. After 5 min add 2 ml of 1% Na_2CO_3 solution and note any change in colour. This test is answered by tyrosine, histidine and tryptophan.

Ehrlich's Test

This test is specific for indole groups present in tryptophan.

Reagent

10% dimethyl amino benzaldehyde in 10% HCl.

Procedure

To 1 ml of the test solution add 1 ml of the reagent and mix well. Appearance of a red colour indicates the presence of tryptophan.

Sakaguchi Test

This test is specific for guanidium compounds and the amino acid that contains this group is arginine.

Reagents

α-napthol: 1% solution in alcohol

Bromine water: Add a few drops of bromine to 100 ml of water, in a fume chamber.

Procedure

Mix 1 ml of NaOH with 3 ml of the test solution and add 2 drops of α-napthol solution. Mix thoroughly and add 4 – 5 drops of bromine water. Watch for the formation of red colour which indicates the presence of arginine.

Test for Sulphur Containing Amino Acids

Sulphur containing amino acids, such as cysteine and cystine upon boiling with sodium hydroxide yield sodium sulphide. This is due to the partial conversion of the organic sulphur to inorganic sulphide, which can be detected by precipitating it to lead sulphide, using lead acetate solution.

Reagents

Sodium hydroxide: 40%

Sodium plumbate reagent: 5 ml of 0.1 N NaOH with 2 ml of 0.1 M lead acetate and heated till the white precipitate dissolves.

Procedure

To 2 ml of the test solution add 0.5 ml of NaOH and boil in a water bath for 2 min and cool. To this add 0.5 ml of sodium plumbate reagent, a black precipitate is formed by sulphur containing amino acids.

Biuret Test

This test is general for compounds having a peptide bond. Alkaline copper sulphate reacts with compounds containing two or more peptide bonds to give a violet or pinkish coloured product which is due to formation of coordination complex of cupric ions with unshared electron pairs of peptide nitrogen and oxygen of water.

Reagents

Copper sulphate: 1% (w/v) of $CuSO_4 \cdot 5H_2O$

Sodium hydroxide: 40% solution.

Procedure

To 1 ml of the test solution add 0.5 ml of NaOH and mix well. Then add 2-5 drops of copper sulphate solution and observe for the formation of pink or violet colour. This shows the presence of peptides or proteins in the sample.

5.3 ESTIMATION OF GLYCINE-SORENSON FORMAL TITRATION

The carbonyl group of amino acids cannot be accurately estimated by titration against alkali. But it reacts with basic amino acids to form zwitterions that may not be decomposed at the end point of the alkali solution indicator. If amino acids are titrated with an excess of formaldehyde it blocks the amino group permitting the carbonyl group to exert the maximum activity.

Reagents

Standard oxalic acid (0.1 N): Weigh 0.63 g of oxalic acid and make up to 100 ml with water.

Sodium hydroxide solution (0.1 N):

Standard glycine solution: Dissolve 75 g of glycine in 100 ml of water.

Formaldehyde solution.

Procedure

Pipette out 5 ml of oxalic acid into a clean conical flask and titrate against the sodium hydroxide taken in the burette. Phenophthalein is used as the indicator. The end point is the appearance of a pale permanent pink colour. From the titre value obtained the normality of sodium hydroxide is standardized.

Pipette out 5 ml of glycine and 5 ml of formaldehyde shake well for 5 min. Titrate this against the standard sodium hydroxide using phenophthalein as indicator. Keep a blank containing 5 ml of water and 5 ml of formaldehyde. From the titre value the amount of glycine can be calculated.

Reference

Sorenson, S.P.L. (1907), Biochem. Z. 7:45.

5.4 ESTIMATION OF AMINO ACID BY NINHYDRIN

Ninhydrin, triketohydrindene hydrochloride reacts with amino acids to give a purple coloured complex (Ruhemann's purple) with an absorption maximum at 570 nm. It oxidizes the amino acids to aldehyde, releasing carbon-di-oxide and ammonia. During the course of the reaction, ninhydrin gets reduced to hydrindantin. The hydrindantin formed condenses with ninhydrin in the presence of ammonia to yield a purple coloured complex.

Reagents

1. **Citrate buffer 0.2M, pH 5.0:** Dissolve 10.51 g of citric acid and 14.71g of sodium citrate in 150 ml of distilled water. Adjust the pH of the solution to 5.0 by adding sodium hydroxide solution (1 N) and make up to 250 ml with distilled water.
2. **Ninhydrin reagent:** Dissolve 0.4 g of stannous chloride in 250 ml of 0.2 M citrate buffer, pH 5.0. To this solution add 50 ml of ethylene glycol monomethyl ether containing 1 g of ninhydrin and mix the contents.
3. **Diluent solution:** Mix n-propyl alcohol and distilled water in equal proportions (1:1).
4. **Standard solution:** Dissolve 100 mg of glycine in 100 ml of water. Working solutions can be made by diluting suitable volumes of standards.

Procedure

To 0.1 ml of the sample and standard amino acid in the range of 0.1 – 1.0 μ mole add 2.0 ml of ninhydrin reagent. Heat the test tubes covered with glass marbles, in a boiling water bath for15 min. Cool the tubes to room temperature and add 7 ml of diluent solution and mix well. Measure the colour developed against the reagent blank at 570 nm.

Reference

Cocking, E.C. and Yemm, E.W. (1954), Process Biochem. 19: 58.

5.5 ESTIMATION OF PHENYLALANINE

The colorimetric estimation of phenylalanine in protein hydrolysate is based on two fundamental principles: complete suppression of the interference from tyrosine and histidine, and the quantitative nitration and reduction of phenylalanine to the violet coloured ammonium salt of diaci-o-dinitrobenzoic acid.

Reagents

1. **Nitration mixture:** Dissolve 10 g of potassium nitrate dissolved in 100 ml of conc. sulphuric acid by shaking at room temperature.
2. **Acid permanganate solution:** Dissolve 5 g of potassium permanganate in 100 ml of 10% sulphuric acid.
3. **Hydroxylamine – ammonium sulphate mixture:** Dissolve 15 g of hydroxylamine hydrochloride and 20 g of ammonium sulphate in 100 ml of distilled water.
4. **Phenylalanine standard:** Dissolve 100 mg of dl-phenyalanine in 100 ml of distilled water.

Procedure

Transfer an aliquot of the protein hydrolysate containing 25 to 50 mg nitrogen in to a volumetric flask and make up the volume to 25 ml with distilled water and adjust the pH to 4. Prepare a column containing 10 g of permutit (synthetic zeolite) contained in a 150 mm calcium chloride tube plugged with coarse glass wool and fitted with a short piece of rubber tubing and a pinch-cock to regulate the flow. Pass the sample through the column at the rate of 1 drop per second, this removes histidine from the solution. Test the solution for histidine, to make sure it is removed. Now take 3 and 5 ml of samples in porcelain evaporating dish and add excess of acid permanganate solution, this destroys the tyrosine in the hydrolysate. Place the dish in an oven at 120°C. Transfer aliquots of 1and 2 ml of phenylalanine standard into a porcelain dish and place this in an oven to evaporate the sample. Take care not to heat longer than 18 to 24 hours, since prolonged heating will result in the destruction of phenylalanine. To each of the residue of the sample add 2 ml of nitration mixture in such a manner that the residue is washed down from the sides. Nitration is completed by heating the samples on a hot air bath for 30 min. Transfer the contents of the evaporating dish into a 50 ml volumetric flask washing the dish with 10 ml distilled water. Add 5 ml of hydroxylamine – ammonium sulphate reagent to all the samples, after 5 min add 20 ml of 20% sodium hydroxide solution slowly. Adjust the volume to 50 ml with distilled water and cool the flasks in an ice bath for 10 min and read the colour at 520 nm.

Reference

Kapeller-Adler, R. (1932), Biochem. Z. 252: 185.

5.6 ESTIMATION OF TRYPTOPHAN

Tryptophan condenses with p – dimethylaminobenzaldehyde (Ehrlich's reagent) under acidic conditions to a colourless condensation product, which upon further oxidation with sodium nitrate yields a blue colour that has an absorption maximum at 590 nm.

Reagents

1. Ehrlich's reagent: p- dimethylaminobenzaldehyde (3% w/v) in 2 N HCl
2. Sulphuric acid 23.7 N
3. Sodium nitrite 0.045% in distilled water.
4. Standard tryptophan: Transfer 100 mg of tryptophan to a standard flask. Suspend the amino acid in 20 – 30 ml of distilled water. Dissolve the amino acid by adding 2 ml of 5N HCl and make up to 100 ml with distilled water. The working standard is prepared by diluting 10 ml of the stock to 100 ml.

Procedure

To 1 ml of the sample add 1 ml of Ehrlich reagent and 8 ml of sulphuric acid and mix well. Incubate the tubes at room temperature, in the dark for 60 min. After incubation, add 0.1 ml of sodium nitrite solution and mix well. Allow the tubes to stand for 30 min at room temperature for colour development. The colour developed is read at 590 nm against a reagent blank. Standard tryptophan in the range of 10 – 100 µg is treated in the same manner.

References

1. *Spies, J.R., and Chambers, D.C. (1948), Anal Chem. 20: 30.*
2. *Spies, J.R., and Chambers, D.C. (1949), Anal Chem. 21: 1249.*

5.7 ESTIMATION OF TYROSINE

The phenolic group of tyrosine reacts with nitrosonaphthol under acidic conditions to yields a red coloured complex, which has an absorbance maximum at 520 nm. The colour developed is stable and the method is sensitive, reliable and can be directly used for the estimation of tyrosine.

Reagents

1. Nitrosonaphthol reagent: 0.15% w/v in 0.1 N sodium hydroxide solution.
2. Acid – base reagent: Mix, equal volumes of nitric acid (0.025 N) and sodium hydroxide (0.3N).
3. Standard solution: Weigh 150 mg of tyrosine and made up to 100 ml with water. Dissolve tyrosine by adding 5 N HCl. Working standard is prepared by diluting 10 ml of stock to 100 ml with water.

Procedure

To 1 ml of the sample add 1 ml of nitrosonaphthol reagent and 2 ml of acid – base reagent and mix. Heat the tubes in a boiling water bath for 10 min. Cool the tubes to room temperature and add 4 ml of conc. H_2SO_4 drop wise. Measure the red colour developed against a reagent blank at 520 nm. Standard solutions in the range 10 – 150 µg is processed in the same manner.

Reference

Uehara, K., Mannen, S. and Kishida, K.J. (1970), Biochem. 68: 119.

5.8 ESTIMATION OF TOTAL SERUM PROTEINS AND ALBUMIN/GLOBULIN RATION BY BIURET METHOD

Globulins are precipitated by a mixed sodium sulphate and sulphite solution in proportions of 3 : 1. In the mixed solution there is no appreciable denaturing effect.

Reagents

1. Sulphate - sulphite solution: Dissolve 208 g of sodium sulphate (anhydrous) and 70 g of sodium sulphite (anhydrous) in 900 ml of water and add 2 ml of concentrated sulphuric acid. Transfer to a one litre volumetric flask and make up with water. The pH should be above 7.0.
2. Stock Biuret reagent: Dissolve 45 g of Rochelles salt in about 400 ml of 0.2 N Sodium hydroxide and add 15 g of copper sulphate stirring continuously until solution is complete. Add 5 g of potassium iodide and make up to a litre with 0.2 N sodium hydroxide.
3. Biuret solution for use: Dilute 200 ml of stock reagent to a litre with 0.2 N sodium hydroxide containing 5 g of potassium iodide per litre.
4. Tartarate – iodide solution: Dissolve 9 g of Rochelle salt in 0.2N sodium hydroxide containing 5 g of potassium iodide per litre.
5. Standard protein solution: Dissolve 500 mg of Bovine serum albumin in 100 ml distilled water.

Procedure

To 4 ml of serum add 5.6 ml of sodium sulphate – sulphite reagent and 3 ml of ether. Centrifuge the tubes for 8 min. Remove the upper layer carefully for the estimation of albumin. Mix 2 ml of this with 3 ml of Biuret reagent. A serum blank containing 2 ml of sodium sulphate – sulphite reagent should be read simultaneously. To another aliquot of 1 ml serum add 3 ml of Biuret reagent this would give the estimate of total proteins. The standard protein solution in the range of 1000 – 4000 µg and processed simultaneously. Shake well and incubate for 10 min at 37°C and read at 640 nm.

Calculation

Since total protein and albumin levels are estimated in this procedure the level of globulin is calculated as

Total protein – albumin = globulin

and from this the albumin/globulin ratio is calculated.

Reference

Gornall, A.G., Bardawill, C.S. and David, M.M.J. (1949), Biol Chem. 177: 751.

5.9 ESTIMATION OF PROTEINS

5.9.1 Lowry's Method

The blue colour developed is a result of reaction of the peptide bonds of the protein with cupric copper under alkaline conditions and the reduction of phosphomolybdic acid by tyrosine and tryptophan residues of the protein.

Reagents

1. Lowry's reagent:
   Solution A: 2% sodium carbonate in 0.1 N sodium hydroxide solution.
   Solution B: 0.5% copper sulphate in 1% sodium potassium tartarate.
   50 ml of solution A is mixed with 1 ml of solution B just before use.
2. Folin's phenol reagent
3. Standard bovine serum albumin (BSA).
   100 mg of BSA is dissolved in 100 ml of water. 10 ml of the stock is diluted to 100 ml to get a working standard containing 100 µg/ml.

Procedure

Dilute 0.1 ml of sample to 10 ml and take 0.1 ml of the diluted sample for estimation. Pipette aliquots of standard (20 – 100 µg) into test tubes and make up to 1 ml with water. Blank contains 1 ml water. Add 5 ml of alkaline copper reagent to all the tubes and leave them at room temperature for 10 min. Add 0.5 ml of

Folin's phenol reagent and read the absorption of the blue colour developed at 640 nm against a reagent blank after 20 min. Protein values are expressed as g/dl.

Reference

Lowry, O.H., Rosenbrough, N,J., Faren, A.L. and Randall, R.J.J. (1951), Biol Chem. 193: 265.

5.9.2 Bradford Method

This is a rapid, simple and sensitive method for estimation of proteins in a sample extract. The colour development is virtually complete in 2 minutes and the colour is stable for about one hour. Unlike Lowry's method, metal ions such as NH_4^+, Na^+, K^+, phenols and carbohydrates such as sucrose do not interfere in this assay. The procedure is based on interaction of a dye, Coomassie Brilliant Blue, with proteins. The unbound dye has an absorbance maxima at 465 nm. However, on interaction with proteins the dye turns blue and its absorbance maxima is displaced to 595 nm. Thus from the absorbance at 595 nm the amount of protein in a sample solution can be quantitatively estimated. However, as in Lowry's procedure detergents such as SDS, Triton X-100, etc., interfere in estimation of proteins by this method.

Reagents

1. Phosphate buffer: 0.1 M, (pH 7.5)
2. Bradford reagent: Dissolve 100 mg of Coomassie Brilliant Blue G 250 in 50 ml of ethanol, add 100 ml of 85% of phosphoric acid and make the volume to one litre with water.
3. Standard protein solution: Dissolve 5 mg of Bovine serum albumin in 50 ml of 0.1M phosphate buffer. This solution contains 100 µg protein/ml.

Procedure

Dilute 0.1ml of sample to 1 ml with phosphate buffer and take 0.1 ml is for estimation. Pipette out aliquots of standard (20 – 100 µg) into test tubes and make up to 1 ml with water. Blank contains 1 ml of phosphate buffer. Add 5 ml of Bradford's reagent to all the tubes and mix thoroughly. Read the absorbance at 595 nm against a reagent blank and the amount of protein is calculated from the standard graph.

Reference

Bradford, M.M. (1976), Anal. Biochem. 72: 248.

5.9.3 BCA Assay Method

This assay measures the formation of Cu^{+1} from Cu^{+2} by the Biuret complex in alkaline solutions of protein. It was originally thought that the mechanism of the assay was the same as in the Lowry's assay, but it has since been determined that there are two distinct reactions that take place with copper ions unique to the BCA assay. The first reaction occurs at lower temperatures and is the result of the interaction of copper and BCA with the following residues: cysteine, cystine, tryptophan and tyrosine. At elevated temperatures it has been shown that the peptide bond itself is responsible for color development. This is why performing the assay at 37°C or 60°C increases the sensitivity and reduces the variation in the response of the assay to protein composition

Reagents

Bicinchoninic Acid Solution: Commercially available (e.g. Sigma B 9643). Contains BCA, sodium carbonate, sodium tartarate and sodium bicarbonate in 0.1 M NaOH (pH 11.25)

Copper Sulphate Solution: 4% $CuSO_4 . 5H_2O$

Prepare the required amount of protein determination reagent by adding 1 volume copper sulfate solution to 50 volumes of bicinchoninic acid solution

Procedure

Pipette samples and standards of bovine serum albumin in the range of 5 to 100 μg in 0.1 ml total volume. Add 2.0 ml of the protein determination reagent to each tube and vortex. Incubate the tubes at 37°C for 30 min (alternatively 2 h at room temperature, or 15 min at 60°C). Cool the tubes to room temperature and determine the absorbance at 562 nm. After cooling to room temperature, the blank continues to increase in absorbance at approximately 2.3% per 10 min (if you use 60°C incubation temperature, the absorbance will not increase appreciably). Because of the small volume in each tube, cuvettes with an approximate 1 ml capacity should be used for the absorbance measurements.

References

1. *P.K. Smith et al. (1985), Anal. Biochem. 150: 76.*
2. *K.J. Wiechelman et al. (1988), Anal. Biochem. 175: 231 (This describes the mechanism of the assay)*

5.9.4 Ultraviolet Absorption Method of Protein Estimation

This method is useful only with comparatively pure solution. Most proteins exhibit a distinct absorption peak in the ultraviolet range at 280 nm and this is due primarily to the presence of tyrosine and tryptophan. The specific extinction coefficient of proteins vary in the range of 6 – 62, depending on their tyrosine and tryptophan content. On an average most proteins have a value close to 10. This is used as the basis of estimating protein concentration.

Protein concentration (mg/ml) = 1.55 × O.D 280 – 0.76 × O.D. 260

The experimental procedure is simple: The optical density of the sample is read both at 260 and 280 nm in a spectrophotometer.

5.10 ESTIMATION OF GLYCOSYLATED HEMOGLOBIN

Glycosylated haemoglobin (HbA_1c) is estimated best in the hemolysate obtained during the isolated of erythrocyte membrane.

Reagents

1. Sodium chloride – 0.9%
2. Oxalic acid – 0.3 M
3. TCA – 40%
4. TBA – 0.05 M
5. Standard fructose in the range of 10 – 40 μg

Procedure

Mix 0.2 ml of hemolysate with 1.8 ml of 0.3 M oxalic acid and hydrolyze the mixture for two hours, cool and add 1.0 ml of 40% TCA. Centrifuge the mixture at 1,400 xg for 20 min and treat 1.5 ml of the supernatant with 0.5 ml of 0.05 M TBA. Incubate the tubes at 37°C for 40 min. read the absorption of the colour developed at 440 nm. Process standard fructose in the range of 10 – 40 μg similarly. Values are expressed as percent of glycosylated hemoglobin present in 100 ml of blood.

Reference

Wang, J. and Yang, G. (1982), Human Yixylynan Xuebao, 325 – 328.

5.11 ESTIMATION OF GLYCOSYLATED PLASMA PROTEINS

The plasma proteins are precipitated, hydrolyzed and the hexoses released are measured by the colour reaction with TBA.

Reagents

1. Sodium chloride – 0.5N
2. TCA – 2.5M
3. Oxalic acid – 1M

4. Saturated solution of TBA
5. Standard fructose in the range of 10 – 40 μg.

Procedure

Dilute 0.5 ml of plasma with 4.5 ml of ice cold sodium chloride and precipitate the proteins by the addition of 0.5 ml of ice cold TCA, centrifuge at 1,500 xg for 15 min at 4°C. Discard the supernatant and allow the sediment to drain for 5 min. Suspend the protein in 4.5 ml of ice - cold sodium chloride and sonicate for 15 min till a translucent to clear stable preparation is obtained. Mix 2.0 ml of the protein solution and 1.0 ml of oxalic acid in a 15 ml glass centrifuge tube and heat in a boiling water bath for 4 – 5 hrs. Place glass marbles on top of the tubes to reduce loss through evaporation. Cool the tubes cooled in ice for 10 min, add 1ml of TCA and vortex for 30 sec. Centrifuge the sample at 1,500 xg for 15 min at 4°C. To 3 ml of the supernatant add 0.75 ml of saturated TBA and incubate for 30 min at 40°C. Read the absorbance of the colour developed against an appropriate blank at 443 nm allowing 15 min for equilibration to room temperature. Process standard fructose in the range of 10 – 40 μg similarly. Before the hydrolysis, determine the protein concentration in the sonicated samples by method of Lowry.

Glycosylated plasma protein is expressed as μg hexose liberated/mg protein.

Reference

Ma A., Naughton M.A., Cameron D.P. (1981), Clin. Chim. Acta. 115: 111.

5.12 PAPER CHROMATOGRAPHIC SEPARATION OF AMINO ACIDS

This is the simplest method for separation of amino acids from a mixture.

Solvents System

Butanol-acetic acid-water 4:1:5 (v/v)

Phenol-water 80: 20 (v/v)

The solvents are mixed in the right proportion in a separating funnel and shaken well. The solution appears to be separated into two phases, the aqueous phase at the bottom and the organic phase on the top.

Spray Reagent

Ninhydrin reagent: Dissolve 200 mg of ninhydrin in 100 ml of 95 % alcohol.

Procedure

Take Whatman No. 1 filter paper sheet of appropriate size. Draw a line at the base 2 cm from the bottom and mark spots on the base line with a distance of 2.5 cm between the adjacent spots.

Apply 2 μl of the sample using a fine capillary tube. The diameter of the spots should be kept as small as possible. Allow the spots to dry, now roll the paper into a cylinder, such that the spots face the outer side. Place the paper in the cylinder containing the running aqueous phase of the solvent system. This chamber should be previously saturated with the organic phase. The saturation should be complete in order to have an effective run.

Allow the solvent to rise to three-fourths of the paper. Remove it and air dry. Spray the paper with the detecting solution and warm the chromatogram in an oven at 80°C, the spots of amino acids become visible.

The amino acids so separated can be quantified by eluting the spots. Cut the spots and put them into a test tube containing 5 ml of methanol and read colorimetrically.

Reference

Consden, R., Gordon, A.H. and Martin, A.J.P. (1944), Biochem. J., 38: 224.

Chapter 6

Lipids

6.1 QUALITATIVE TESTS FOR LIPIDS

1. Solubility Test

Lipids show little or no solubility in water and they form emulsions in water. Triglycerides are soluble in chloroform, diethyl ether and benzene.

Reagents

1. Fat sample
2. Solvents (water, acetone, ethanol, diethyl ether, chloroform)

Procedure

Take a small amount of lipid sample and test its solubility in the above mentioned solvents and record the observations.

2. Specific Gravity of Fats and Oils

The specific gravity of liquids is the weight of the given volume of liquid at the specific temperature, compared to the weight of an equal volume of water at the same temperature, all weights measured in presence of air.

Weigh the specific gravity bottle when empty = W_B

Fill the bottle completely with the liquid and weigh = W_L

After cleaning, the bottle is filled completely distilled water and weigh = W_W

The temperature of the liquid is noted.

$$\textbf{\textit{Specific gravity}} = \frac{W_L - W_B}{W_W - W_B}$$

The temperature at which the specific gravity is taken should be specified.

3. Acrolein Test for Glycerol

Reagents

1. Lipid samples
2. Anhydrous potassium bisulphate

Procedure

Take a small amount of potassium bisulphate ($KHSO_4$) in a test tube, add a few drops of test solution, and a further layer of ($KHSO_4$) and heat slowly. A pungent odour is liberated. This is due to acrolein which is derived from glycerol.

4. Qualitative Test for the Presence of Fatty Acids by Titrimetric Method

The non-esterified fatty acids are determined by titrating with alkali using phenolphthalein as indicator.

Reagents

1. Lipid samples (dissolved in 50% ethanol)
2. Phenophthalein indicator – 1%
3. 0.1 N NaOH

Procedure

In an Erlenmeyer flask take 10 ml of 0.1N NaOH and add a drop of phenolphthalein solution which will give permanent pink colour. To this add the test solution drop by drop with constant shaking. The disappearance of the pink colour shows the presence of fatty acids.

4. Test for Unsaturation of Fatty Acids

This is a semiquantitative estimate of the degree of unsaturation of lipid samples which can be estimated by the fact that halogens readily add across the double bonds.

Reagents

1. Lipid solution
2. Bromine water: Add 5 ml of bromine to 100 ml of water and keep in a dark bottle.

Procedure

To approximately 5 ml of the test solution slowly add bromine water drop-wise with shaking. Keep on adding bromine water till it fails to get decolourised and retains its colour. From the volume of bromine water added the extent of unsaturation can be determined.

5. Qualitative Estimation of Cholesterol by Liebermann-Burchard Method

Reagents

1. Test solution
2. Acetic anhydride-sulphuric acid mixture: Prepare by mixing acetic anhydride and H_2SO_4 (30:1 *v/v*) just before use.
3. Cholorform

Procedure

Take a few drops of test solution in a test tube and add a few drops of the reagent, cover the test tubes and leave them in dark for 15 min. A green colour appears showing the presence of cholesterol.

This method can also be used to estimate cholesterol quantitatively.

6.2 EXTRACTION OF TOTAL LIPIDS

1. Bloor's Method

A mixture of ethyl ether and ethanol (3:1) is used. Grind the tissue (1g wet weight) well in a mortar with about 10 ml of the solvent mixture and the extract is transferred to a separatory funnel. To this add 2 ml of 0.05 M KCl solution and shake well. Two layers will separate. While the water in KCl solution helps in layer separation, the salt prevents any emulsification.

2. Folch's Method

Homogenize the tissue in a Potter-Elvehjem homogenizer. Filter the contents, carefully scrap off the residue and homogenize with chloroform. Allow the solvent to evaporate below 60°C and redissolve lipid extract in a known volume of chloroform – methanol mixture (2 : 1 *v/v*). To this add 1.0 ml of 0.1 M KCl. The contents are shaken well and centrifuged. Wash the aqueous phase again with KCl – methanol – chloroform mixture (10 : 10 : 1 *v/v*), 3 times and each time discard the upper phase.

6.3 ACID VALUE OF FATS AND OILS

The acidity of fats and oils is expressed as acid value or number, which is defined as mg KOH required to neutralize the free fatty acids present in 1 g of fat and oil. The amount of free fatty acids present in the fat is a useful parameter which gives information regarding the age and extent of its deterioration. Since fats become rancid and rancidity occurs because of chemical and enzymatic hydrolysis of fats into free acids and glycerol, the amount of free acid is estimated titrimetrically using KOH.

Reagents

1. 1% phenolphthalein indicator
2. 0.1 N potassium hydroxide solution.
3. 95% ethanol

Procedure

Take the sample (5 g fat) in a conical flask and add 25 ml of ethanol to dissolve it. Shake well and add few drops of phenolphthalein indicator to the sample. Titrate this solution against standard 0.1 N KOH (which is initially standardized with 0.1N oxalic acid). The end point is the appearance of faint pink colour which is persistent for 20 – 30 sec. Note the volume of KOH used. A blank (only the fat solvent) is also titrated with KOH.

Calculations

N KOH solution used by blank = x ml
N KOH solution used by test sample = y ml
Titre value for the sample = $(y-x)$ ml

$$\text{Acid value (mg KOH/g fat)} = \frac{\text{Titre value} \times \text{Normality of KOH} \times 56.1}{\text{Weight of sample (g)}}$$

Reference

Official and Tentative Methods of the American Oil Chemist's Society, AOCS, 3rd ed. Illnois, USA, (1981).

6.4 IODINE VALUE OF FATS AND OILS

The fatty acids in triglycerides may be saturated or unsaturated. The amount of unsaturation is determined by measuring the amount of iodine taken up by the unsaturated fatty acids.

Reagents

1. Wij's solution: Dissolve 8.5 g of iodine and 7.8 g of iodine trichloride separately in 450 ml acetic acid each. Mix both the solutions and make the volume up to 1 L.
2. 0.1 N sodium thiosulphate (the normality checked by titrating with 0.1 N potassium dichromate).
3. 10% potassium iodide
4. 1% starch indicator

Procedure

Take 10 ml of fat solution into iodine flask and add 25 ml of Wij's solution. Shake thoroughly and allow it to stand in dark for 1 h. Prepare a blank similarly but without the test solution. After the reaction time of 1 h, rinse the stopper and the neck of the iodine flask with 50 ml of water and add 10 ml of potassium iodide solution. Titrate the liberated iodine with standard sodium thiosulphate solution till the content in the flask become straw yellow in colour. Add a few drops of starch solution and the contents of the flask turns blue in colour. Continue to titrate it further with sodium thiosulphate solution till the blue colour disappears.

Calculation

Volume of 0.1 N sodium thiosuphate used by the blank = x ml

Volume of 0.1 N sodium thiosuphate used by the test = y ml

$$\text{Iodine number} = (x - y) \times \frac{12.7}{1000} \times \frac{100}{\text{Wt of sample (g)}}$$

Reference

Official and Tentative Methods of the American Oil Chemist's Society, AOCS, 3rd ed. Illnois, USA, (1981).

6.5 SAPONIFICATION VALUE OF FATS AND OILS

When the oil is heated with KOH (or alkali) it is saponified (hydrolyzed) and releases fatty acids and glycerol. The fatty acids neutralize the sodium hydroxide and the amount of alkali that has been used for saponificaiton. Each mole of triglyceride uses 3 moles of KOH for saponification. The saponificaiton value is defined as mg of KOH required to saponify 1 g of the given fat.

Reagents

1. 0.5 N alcoholic KOH: Dissolve KOH in ethanol to give a 0.5 N solution.
2. 1 % Phenolphthalein indicator

3. 0.5 N HCl
4. Fat solvent: A mixture of 95% ethanol and ether (1:1 *v/v*)

Procedure

Weigh accurately 1 g of fat in a conical flask and dissolve it in about 3 ml of the fat solvent. Add 25 ml of 0.5 N alcoholic KOH, attach a reflux condenser to it and reflux the contents on boiling water bath for 30 min. Cool to room temperature and add a few drops of phenolphthalein into the flask. Titrate the contents of the flask with 0.5 N HCl till the pink colour disappears. Similarly run a blank by refluxing 25 ml of 0.5 N alcoholic KOH without the fat sample.

Calculation

0.5 N KOH (blank) = x ml

0.5 N KOH used by the test sample = y ml

Titre value = $(x - y)$ ml

$$\text{Saponification value} = \frac{28.05 \times \text{titer value}}{\text{Weight of sample (g)}}$$

Reference

Official and Tentative Methods of the American Oil Chemist's Society, AOCS, 3rd ed. Illnois, USA, (1981).

6.6 ESTIMATION OF TRIGLYCERIDES IN PLASMA

Lipids are extracted with chloroform-methanol mixture (2:1 *v/v*) phospholipids are adsorbed on silicic acid and the triglycerides remaining in the solution are saponified with alcoholic KOH. The liberated glycerols are oxidized by peroidate to formaldehyde and the excess oxidizing power is destroyed by reaction with sodium arsenite. The formaldehyde formed is determined by the addition of chromotropic acid.

Reagents

1. Chloroform – methanol mixture (2:1 *v/v*)
2. Saturated NaCl
3. Activated silicic acid
4. 0.4% alcoholic KOH
5. 0.2% sulphuric acid
6. 0.1 M sodium metaperiodate
7. 0.5 M sodium meta arsenite

8. Chromotropic acid reagent – Dissolve 1.10g of sodium salt of chromotropic acid in 100 ml of distilled water. Then add 450 ml of dilute acid solution (H_2SO_4 : H_2O 2:1 *v/v*) and stir well.
9. 1.0% thiourea solution made in water
10. Prepare Tripalmitin standard by dissolving in chloroform. The working standard contains 0.1 mg/ml.

Procedure

Mix 0.2 ml of plasma thoroughly with chloroform : methanol mixture and leave aside for 30 min. Centrifuge and mix 0.4 ml of the supernatant (lipid extract) with 8 ml of sat. NaCl and shake vigorously. Allow the contents to settle for 1 hr, centrifuge and discard the supernatant saline-methanol phase. Filter the washed chloroform phase into a dry tube. Add 200 mg of activated silicic acid to the chloroform phase, shake vigorously allow to stand for 30 min and centrifuge. Remove 0.5 ml of the supernatant as well as tripalmitin standards and evaporate to dryness. Add 0.5 ml of alc. KOH solution to the test, standard and blank and saponify the mixture in a water bath at 60 – 70°C for 20 min. Add 0.5 ml of 0.2 N sulphuric acid and heat in a boiling water bath for 10 min. Cool the tubes, add 0.1 ml of sodium meta periodate and allow to stand for 10 min. Add 0.1 ml of sodium arsenite to reduce the excess periodate. Add 0.5 ml of chromotropic acid reagent, whilst keeping the tubes in an ice – bath, mix thoroughly and keep in a boiling water-bath for 30 min. Cool and add 0.5 ml of thiourea, the colour developed is read at 540 nm. Triglycerides are expressed as mg/dl plasma.

Reference

Rice, E.W. (1970), Standard methods in clinical chemistry, Vol 6, ed. Roderice, P. and Mc Donald, Academic Press, 215.

6.7 ESTIMATION OF PHOSPHOLIPIDS IN PLASMA

Proteins are precipitated by trichoroacetic acid and the precipitate, which contains the phospholipids, is digested with perchloric acid. The inorganic phosphorus is determined.

Reagents

1. TCA – 10%
2. Ammonium molybdate – 3%
3. Ascorbic acid – 3%
4. Standard phosphate solution:

Dilute 0.1 ml of plasma to 2.0 ml with distilled water and add 2.0 ml of 10% TCA. Precipitate the proteins by centrifugation. Discard the supernatant and add 1.0 ml of perchloric acid (10%) to the residue and digest on a sand bath till the solution becomes colourless. Cool the solution and make up to 5.0 ml with distilled water. Mix standard phosphate solutions and blank containing water with 0.8 ml perchloric acid and make up the final volume to 5 ml with distilled water. Add 0.5 ml each of ammonium molybdate and ascorbic acid and keep the mixture in a boiling water bath for 6 min. The blue colour developed was read at 710 nm.

Phospholipids were expressed as mg/dl plasma after multiplication by a factor 25 to give the phospholipid content.

Reference

Rouser, G., Fleischer, S. and Yammoto, A. (1970), Lipids, 5: 494.

6.8 ESTIMATION OF FREE FATTY ACID (FFA) IN PLASMA

Free fatty acids are extracted with chloroform: heptane : methanol solvent, silicic acid removes phospholipids from interfering with the estimation of free fatty acids. FFA forms a complex FFA-Cu, and the copper is estimated with dithiocarbamate.

Reagents

1. Chloroform : heptane : methanol solvent (CHM), 200 : 150 : 7 (*v/v*)
2. Copper nitrate – triethanolamine (Cu – TEA) solution: Mix 50 ml of 1.0 M copper nitrate and 50 ml of 1.2 M triethanolamine with 33 g of NaCl. Adjust the pH to 8.1.
3. Colour reagent – 0.1% diethyl dithiocarbamate in n-butanol
4. Activated silicic acid
5. Standard palmitic acid containing 200 μg/ml dissolved in chloroform.

Procedure

Mix 6 ml of CHM solvent with 0.2 ml plasma and shake vigorously. Add 200 mg of activated silicic acid shake well and leave aside for 30 min. Centrifuge and transfer the supernatant to tubes containing 2.0 ml of Cu-TEA solution. Vortex the contents for 20 min and centrifuge, the mixture will separate into two phases. Mix 2.0 ml of the upper layer with 1.0 ml of the colour developing reagent and read at 430 nm. Process standards and blank simultaneously.

Plasma free fatty acids are expressed as mg/dl.

Reference

Hron, W.P. and Menahan, L.A. (1981), J. Lipid Res. 27: 377.

6.9 ESTIMATION OF ESTERIFIED FATTY ACIDS IN PLASMA

Hydroxylamine in alkaline solution reacts with esters of fatty acids to give hydroxamic acids which produces a red to violet colour with ferric chloride.

Reagents

1. Ethanol-ether mixture: Mix three parts of ethanol and one part of ether. Purify the ethanol by refluxing for several hours with 5% sodium hydroxide and 5% zinc dust. Distill and redistill.
2. Hydroxylamine hydrochloride – 14% aqueous solution
3. Ferric chloride – 10% in 0.1N HCl
4. Sodium hydroxide – 14%
5. Hydrochloric acid – 1 volume HCl with 2 volumes water
6. Stock standard – Dissolve 1.18 g of triolein in ethanol-ether mixture and make up to 100 ml.
7. Working standard – Dilute 1 ml of stock with 20 ml of ethanol-ether mixture.

Procedure

Pipette 0.5 ml of sample into 8 ml of ethanol-ether mixture in a 10 ml stoppered tube. Heat to boiling with shaking. Cool and make up to 10 ml with ethanol-ether mixture and Filter. To 3 ml of filtrate add 0.5 ml of 14% hydroxylamine hydrochloride, shake well, add 0.5 ml of 14% sodium hydroxide and shake again. Stopper the tubes and allow to stand at 25°C for twenty minutes. Add 0.6 ml of hydrochloric acid, followed by 0.5 ml of 10% ferric chloride. At the same time put up a standard using 1 ml of working standard and 2 ml of ethanol-ether mixture, treating both as described as test. Read against the blank using a yellow-green filter or at 540 nm.

Reference

Stern, I.S. and Shapiro, B. (1953), J. Clin. Pathol. 6: 158.

6.10 ESTIMATION OF CHOLESTEROL AND THE LIPOPROTEIN FRACTIONS IN PLASMA

Cholesterol estimation in the plasma and in the lipoproteins fractions are estimated by ferric chloride – uranyl acetate method.

Reagents

1. Ferric chloride – Uranyl acetate reagent: Dissolve 500 mg of ferric chloride in 10 ml water, add 3 ml of liquor ammonia and shake well. Wash the

precipitated ferric hydroxide several times with distilled water, dissolve in glacial acetic acid and make up to a litre with glacial acetic acid. Add 100 mg of uranyl acetate, shake well and leave overnight. Store the reagent in a brown bottle.

2. Sulphuric acid – Ferrous sulphate reagent: Dissolve 100 mg of anhydrous ferrous sulphate in 100 ml glacial acetic acid with stirring. Cool to room temperature and make up the solution a litre with conc. H_2SO_4.
3. Standard cholesterol: Dissolve recrystalized cholesterol in glacial acetic acid (AR) 2 mg/ml.

Procedure

Mix 0.05 ml of plasma (or) 0.1 ml of the lipoprotein subfractions with 10 ml of ferric chloride – uranyl acetate reagent. Mix the contents well and allow to stand for 15 min the proteins will precipitate. Centrifuge and mix 3 ml of the clear supernatant with 2.0 ml of sulphuric acid – ferrous sulphate reagent. Mix 0.05 ml of standard cholesterol with 10 ml acetate reagent and treat 3 ml aliquots with 2.0 ml of acid reagent. Blank contains 3 ml Fe_2Cl_3 – uranyl acetate reagent. The colour developed is read at 530 nm after 20 min.

Cholesterol levels are expressed as mg/dl plasma for both total and the different lipoprotein classes.

Lipoproteins are fractionated by dual precipitation technique and cholesterol estimated in the fractions by the above mentioned method.

HDL and its Subfractions

Reagents

1. Heparin – $MnCl_2$ reagent: Add 3.167 mg of $MnCl_2$ to 1.0 ml solution of heparin containing 20,000 units and make up to 8.0 ml with distilled water.
2. Dextran sulphate: Dissolve 121 mg of dextran sulphate (MW 15,000) in 10 ml of 0.89% saline.

Procedure

Add 2 ml of serum to 0.18 ml of heparin – $MnCl_2$ and mix well. Allow the mixture to stand for 30 min and centrifuge in a refrigerated centrifuge at 25,000 rpm maintaining at 10°C for 30 min. The supernatant contains the HDL fraction. Aliquot of 1 ml is used for the fractionation of HDL subfractions HDL_2 and HDL_3. The remaining supernatant is taken for the estimation of cholesterol.

HDL_3 Assay

Mix 1 ml of the clear total HDL supernatant from the above with 0.1 ml of dextran sulphate and allow to stand at room temperature for 20 min to precipitate HDL_2. The supernatant contains the HDL3 fraction. Cholesterol is assayed in the HDL_3 fraction.

LDL and VLDL

Reagents

1. Sodium dodecyl suphate (SDS) – 10% solution in 0.15M NaCl, pH 9.0
   Add 2 ml of plasma to 0.15 ml of SDS. Mix the contents well and incubate at 37°C for 2 hrs. centrifuge the contents in a refrigerated centrifuge at 10,000 rpm for 15 min. VLDL aggregates as a pellicle at the top. The supernatant contains the HDL and LDL fractions. Cholesterol is estimated in this fraction.

References

1. *Parekh, A.C. and Jung, D.H. (1970), Anal. Chem. 42: 1423.*
2. *Wilson, D.E. and Spiger, M.J. (1973), J. Lab. Clin. Med. 82: 473.*
3. *Gidez, Z.I., Miller, G.J., Burstein, M., Seagle, S. and Eder, H.A. (1982), J. Lipid Res., 23: 1206.*
4. *Burstein, M. and Scholnick, H.R. (1972), Life Sci. 11: 117.*

Estimation of Cholesterol by Zak's Method

Cholesterol in acetic acid gives a red colour with ferric chloride and sulphuric acid.

Reagents

1. 0.05% ferric chloride solution
2. Standard cholesterol containing 1 mg/ml.

Procedure

Pipette out standard cholesterol in the range 20 – 100 μg and 0.2 ml of the sample and make up to 5 ml with ferric chloride – acetic acid reagent. Carefully add 3 ml of concentrated sulphuric acid and mix the contents well. Allow to stand for 10 – 20 min a pinkish red colour develops and is read at 540 nm.

Reference

Zak, B., Dickenbaum, R.L., White, E.G., Burnett, H. and Cherney, P.J. (1954), Amer. J. Clim. Pathol. 24: 1307.

6.11 THIN LAYER CHROMATOGRAPHY OF PHOSPHOLIPIDS

Phospholipids are separated by thin layer chromatography using silica gel G.

Reagents

Solvent system: Chloroform: methanol: acetic acid: water (25:15:4:2 v/v).

Procedure

Take glass plates of suitable dimensions, and coat it with silica gel G, alternatively precoated plates can be used. Allow the gel to dry and activate the plate at 80°C in a hot air oven. Then plot the samples at one end of the plate leaving at least 2 cm from the bottom. Carefully plot the samples taking care to not to overload and keep the diameter of spot as small as possible. Carefully keep the plate in the jar containing the running solvent. Develop the plate until the solvent runs three-fourth's. Remove the plate and the spots can be visualized by keeping the plates in an iodine chamber. The lipids react with the iodine to give a brown colour spot on a yellow background. However, the colour developed is not stable.

Two Dimensional Thin Layer Chromatography

The plates run in one direction in a solvent system are run again at 90° to the direction of the first run. This gives a better separation of compounds that have very close R_f values.

The solvent systems commonly employed are:

Chloroform – methanol – water 65:25:4 v/v

Di-isobutylketone – acetic acid – water 80:50:10 v/v

Di-isobutylketone – pyridine – water 25:25:4 v/v

Chloroform – methanol – 28% ammonia 65:35:5 v/v

The plates so run are developed in the same way as mentioned above.

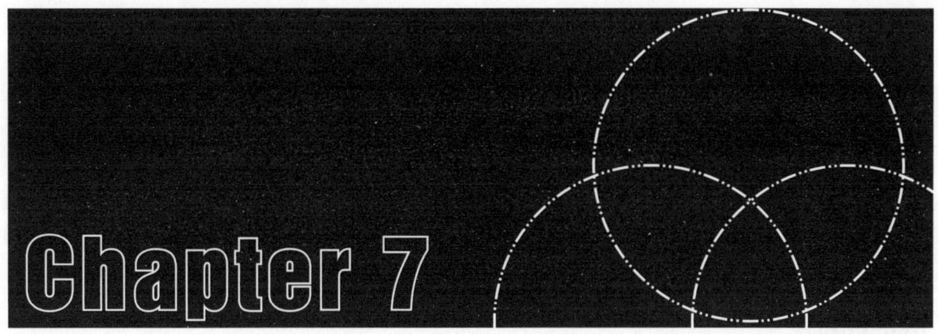

Enzymes

7.1 DETERMINATION OF β-AMYLASE (EC 3.2.1.2) ACTIVITY IN SALIVA

Amylase hydrolyses starch into sugars. The β-amylase is an exohydrolase and it removes two molecules of glucose at a time from the non-reducing end. The products so formed are a disaccharide, β maltose, and limit dextrin. The reducing sugars can be determined quantitatively by their reaction with dinitrosalicylic acid.

Reagents

Buffered starch substrate: Weigh 1g of starch and add a little boiling water to dissolve it and make upto 100 ml with buffer.
Phosphate buffer 0.1M, pH7.0.

Procedure

Pipette 1 ml of appropriately diluted enzyme (salivary amylase 1 in 10 dilution) into two test tubes. Add 1 ml of starch substrate to one tube and label this test and distilled water to the other tube this acts as control. Incubate the tubes at 37°C for 15 min. and arrest the reaction by adding dinitrosalicylic acid (DNS). The estimation of maltose with DNS is performed as outlined in the section.

The activity of the enzyme is expressed in terms of μmoles of maltose produced/ml saliva.

Reference

Winn-Denn, E., David, H., Sigler, G. and Chavez, R. (1988), Clin. Chem., 34: 2005.

7.2 DETERMINATION OF INVERTASE (β-FRUCTO FURANOSIDASE EC 3.2.1.26) ACTIVITY IN YEAST

The enzyme catalyzes the breakdown of sucrose to glucose and fructose. Yeast is a rich source of the enzyme. It is suspended in 0.1M phosphate buffer, pH 7.0 with glucose for one hour with shaking.

Reagents

Substrate: Sucrose 0.3M solution

Citrate-phosphate buffer, 0.2M, pH 4.5.

Sodium hydroxide 1%

Procedure

Pipette out 1.5 ml of citrate – phosphate buffer and add 0.5 ml of the substrate, make upto 2.5 ml. To this add 1 ml of enzyme and incubate the mixture at room temperature for 15 min. Terminate the reaction by adding 0.5 ml of 1% sodium hydroxide and centrifuge to remove sediments if any. Estimate glucose in the supernatant using any one of the methods mention in chapter 4.

The activity of the enzyme is expressed as mg glucose liberated/min.

Reference

Bergmeyer, (1974). Methods of Enzymatic Analysis, Vol. I, 450.

7.3 DETERMINATION OF ALDOLASE ACTIVITY (EC 4.1.2.13)

Aldolase catalyzes the splitting of fructose-1,6-bisphosphate to dihydroxyacetone phosphate and 3-phosphoglyceraldehyde.

Reagents

1. Collidine buffer, 0.1M, pH 7.4. Dissolve 2.42 g of 2.4.6 collidine in water, adjust the pH to 7.4 with HCl and make up to 200 ml with water.
2. Hydrazine, 0.56 M, pH 7.4. Dissolve 14.58 g of hydrazine sulphate in warm water, adjust to pH 7.4 with dilute sodium hydroxide and make to 200 ml with water.
3. Iodoacetate, 0.002 M, pH 7.4. Dissolve 80 mg of monoiodoacetatic acid in water, adjust the pH to 7.4 with dilute sodium hydroxide and make to 200 ml with water.
4. Fructose 1,6-bisphosphate, 0.06 M, pH 7.4. Dissolve 321 mg of the tetrasodium salt in 10 ml water.
5. Trichloroacetic acid – 10%
6. 2,4-dinitrophenylhydrazine – 1% in 2 N HCl.
7. Sodium hydroxide – 0.75%

Procedure

Into two test tubes pipette out 1 ml buffer and 0.25 ml each of hydrazine, iodoacetate and water. To each tube add 1 ml of the sample and to the test add 0.25 ml fructose bisphosphate. Mix the contents well and incubate for one hour at 37°C. Then to both tubes add 3 ml of TCA and to the second tube 0.25 ml of fructose bisphosphate. Mix well; stand for 10 min and centrifuge. To 1 ml of the supernatant add 1 ml of 0.75 % sodium hydroxide, allow to stand for 15 min and add 1 ml of dinitrophenyhydrazine. Place the tubes in a water bath at 37°C for 10 min and make up the volume to 10 ml with the sodium hydroxide solution. Read the test against the control at 520 nm.

Activity is expressed as IU/L. One unit of aldolase activity is defined as that which would split 1 μL of fructose bisphosphate in one hour.

Reference

Bruns, F. (1954), Biochem. Z. 325: 156.

7.4 DETERMINATION OF LIPASE (EC 3.1.1.3) ACTIVITY IN SERUM

Fats are hydrolyzed by lipases and esterases. Lipases hydrolyze esters of long chain fatty acids containing 8 to 18 carbon atoms. Thus as substrate for lipase, esters of long chain fatty acids are preferred. Since esterases may also interfere with the hydrolysis, triolene (olive oil) is hydrolyzed only to a small extent by the esterases.

Reagents

1. Phosphate buffer, pH 7.0. Mix 10 ml of M/3 disodium hydrogen phosphate (47.3 g Na_2HPO_4 per litre) and 30 ml of M/3 potassium dihydrogen phosphate (4.53 g of KH_2PO_4 per litre).
2. Ethanol
3. Sodium hydroxide 0.05 N
4. Olive oil, 50% emulsion. Homogenize a mixture of equal parts of olive oil and a 5% solution of gum
5. Phenophthalein indicator, 1%

Procedure

Pipette 3 ml of distilled water into each of two test tubes and add 1 ml of serum to each. Place one in a water bath at 100°C for five minutes to inactivate the lipase, and then cool. To both tubes add 0.5 ml of the buffer solution and 2 ml of the 50% olive oil emulsion. Shake well and incubate at 37°C for 24 h and add 3 ml of 95% ethanol and 2 drops of phenolphthalein, titrate each with 0.05 N sodium hydroxide to a similar pink colour.

Calculation

Lipase activity per ml of serum = ml titration of the unknown – ml titration of control.

One unit of lipase will produce 1.0 µmole of glycerol/min at pH 7.0 and 37°C.

Reference

Cherry, I.S. and Crandall, L.A. Jr. (1932), Amer. J. Phsiol. 100: 266.

7·5 DETERMINATION OF ISOCITRATE DEHYDROGENASE (EC 1.1.1.42) ACTIVITY

Isocitrate dehydrogenase catalyzes the reversible oxidation of isocitrate to oxoglutarate.

Isocitrate + NADP ⟶ NADPH + oxalosuccinate ⟶ α oxoglutarate

Reagents

1. Sodium chloride – 0.15M
2. Tris buffer, pH 7.5, 0.1M in 0.15 M sodium chloride. Add N hydrochoric acid to 1.214 g of tris (hydroxymethyl) aminomethane to bring the pH to 7.5 and dilute to 100 ml with water.
3. Buffered substrate: Dissolve 1.845 g trisodium isocitrate (70% per dl) in 100 ml of the tris buffer.
4. Nicotinamide-adenine dinucleotide phosphate. Dissolve 10 mg per ml 0.15 M sodium chloride.
5. Reduced nicotinamide-adenine dinucleotide phosphate: 10 mg per 0.15 M sodium chloride.
6. Manganous chloride: 0.03 M. Dissolve 0.597 g $MnCl_2.4H_2O$ in 100 ml of 0.15 M sodium chloride.
7. Standard solution of α oxoglutarate: A stock standard is prepared by dissolving 70 mg per 100 ml. Diluted 1 to 10 before use.
8. Ethylene diaminotetra-acetic acid: Dissolve 5.6 g in 10 ml water.
9. 2,4-dinitrophenylhydrazine: Dissolve 19.8 mg in 100 ml N hydrochloric acid.
10. Sodium hydroxide – 0.4 N

Procedure

For the test measure 0.5 ml of buffer substrate, 0.1 ml NADP, 0.1 ml of manganous chloride and 0.3 ml of 0.15 M sodium chloride into a test tube and for blank 0.5 ml of substrate, 0.1 ml manganous chloride and 0.4 ml of 0.15 M sodium chloride. Maintain the tubes at 37°C for 5 min and then to each tube add 0.2 ml of the sample, mix and incubate for an hour.

Set up standards and reagent blank as follows:

	Reagent blank	*Standard I*	*Standard II*
A-oxoglutaric acid	0	0.5	1.0
MnCl2	0.1	0.1	0.1
NADP	0.1	0.083	0.067
NADPH2	0	0.017	0.033
0.15 M sodium chloride	1.0	0.5	0

After incubation remove the tubes from the bath and to these and to reagent blank and standards add 1 ml of EDTA solution and immediately 1 ml of 2,4 dinitrophenylhydrazine. Allow to stand for 20 min and then add 10 ml of 0.4 N sodium hydroxide. Allow to stand for 15 min and read against a reagent blank at 390 nm. Pipette out 0.5 ml and 1.0 ml of standard –oxoglutaric acid solutions (0.5 – 1.0 ml) and add 0.1 ml of $MgCl_2$ to both the tubes. Add 0.083 ml and 0.067 ml of NADP and 0.017 ml and 0.033 ml of $NADH_2$ to both the standard tubes respectively. Add 0.5 ml of sodium chloride to the standard containing 0.5 ml oxoglutaric acid.

Activity is expressed as IU/L.

One unit is expressed as μmoles of α oxoglutarate formed per litre per min at 37°C.

Reference

Taylor, T.H. and Friedman, M.E. (1960), Clin. Chem. 6: 208.

7.6 DETERMINATION OF LACTATE DEHYDROGENASE (EC 1.1.1.27) ACTIVITY

The enzyme catalyses the reaction

$$\text{Pyruvate} + \text{NADH} + \text{H}^+ \longrightarrow \text{Lactate} + \text{NAD}^+$$

Reagent

1. Phosphate buffer (pH 7.4, 100 mM)
2. Stock pyruvate solution: Dissolve 415 mg of sodium private in 100 ml phosphate buffer.
3. Buffered substrate: Dilute 1 ml of the stock substrate solution to 50 ml with phosphate buffer.
4. Nicotinamide dinucleotide (reduced) solution: Dissolve 10 mg in 1 ml of phosphate buffer

5. 2,4-dinitrophenyhydrazine reagent: Dissolve 40 mg of 2,4-dinitrophenylhydrazine in 8.5 ml of conc. HCl and make up the volume to 100 ml with distilled water.
6. Sodium hydroxide solution – 0.4 N.

Procedure

Take three clean test tubes and label them as Blank (B), Control (C) and Test (T). Dispense the assay reagents and test sample in the following sequence:

Blank (B): Add 1.2 ml of phosphate buffer and 1 ml of dinitrophenylhydrazine solutions.

Control (C): Add 1.0 ml of buffered substrate, 0.2 ml of phosphate buffer and 1 ml of dinitrophenylhydrazine solution.

Test (T): Add 1.0 ml of buffered substrate and 1.0 ml of test sample into the tube and place the contents in a constant temperature water bath at 25°C, for 10 – 15 min. Initiate the reaction by the addition of 0.1 ml of NADH solution. Incubate for exactly 15 min, at 25°C. Remove the tube from the water bath, and add 1 ml of dinitrophenylhydrazine solution and vortex.

Allow the tubes to stand at room temperature for 15 – 20 min. Later add 10 ml of sodium hydroxide to all the tubes and mix. Read the absorbance of the colour developed after 10 min at 510 nm.

The enzyme activity is expressed as IU/L. One unit of the enzyme is that which catalyzes the formation of one μmole of pyruvate /min.

Reference

Cabaud, P.C. and Wroblewski, F. (1958), J. Clin. Pathol. 30: 234.

7.7 DETERMINATION OF MALATE DEHYDROGENASE (EC 1.1.1.37) ACTIVITY

Malate dehydrogenase (MDH) is an enzyme of TCA cycle and catalyses the following reaction:

$$\text{Oxaloacetate} + \text{NADH} + \text{H}^+ \xrightleftharpoons{Mg^{++}} \text{malate} + \text{NAD}^+$$

MDH is an oxidoreductase which utilizes NADH for reduction of oxaloacetate to malate.

Reagents

1. Tris – HCl buffer (pH 8.0): 50 mM containing 1 mM EDTA.
2. NADH solution: Make 6 mM NADH solution in 50 mM Tris-HCl (pH 8.0).

3. Oxaloacetate solution: Prepare 0.3 M solution of oxaloacetate in water and neutralize it by adding solid $NaHCO_3$ till gas bubbles cease to evolve.

Procedure

Take 1.5 ml of buffer and 0.1 ml of the enzyme in a quartz cuvette and adjust the spectrophotometer to 100% transmission at 340 nm. Take out the cuvette and incubate in a water bath at 37°C for about 2 – 3 min. Add 0.05 ml of NADH and mix the contents buy inverting the cuvette and place it immediately in the spectrophotometer and note the initial reading. Record the decrease in absorbance at intervals of 15 sec upto 3 min. This serves as a blank. Start the reaction by adding 0.05 ml of oxaloacetate solution and note the decrease in absorbance upto 3 min at regular intervals for 15 sec. Prepare a standard curve of NADH over a range of 0 – 0.3 μmoles. Make sure that total volume of the solution in the cuvette for preparation of standard curve is the same as during the assay and this is done by adding calculated volume of assay buffer.

One unit of the enzyme is that which catalyzes the formation of one μmole of malate/min.

Reference

Sharpe, D., Wilcock, A. and Goldberg, D. (1973), Clin. Chim. 19: 240.

7.8 DETERMINATION OF SUCCINATE DEHYDROGENASE (EC 1.3.99.1) ACTIVITY

The enzyme catalyses the hydrolysis of succinate to fumarate. The rate of the oxidation reaction is coupled to a redox dye. The dye dicholorophenol indophenol acts as hydrogen acceptor form $FADH_2$ and gets reduced to a colourless form. Thus by following the decrease in the blue colour of the dye, the rate of oxidation of succinate can be followed. To prevent $FADH_2$ from getting oxidized through the electron transport chain, an inhibitor azide is included in the system.

Reagents

1. Potassium phosphate buffer, 0.4 M, pH 7.2
2. Sodium succinate, 0.15 M, pH 7.0.
3. Sodium azide, 0.2 M
4. Dicholorophenol indophenol dye: Prepare the dye solution by dissolving 6 mg/ ml water.

Procedure

Prepare an assay mixture containing 5 ml of buffer, 0.2 ml each of sodium succinate, sodium azide and dye. Measure the absorbance at 600 nm. Now add

0.4 ml of the enzyme extract, mix well and read the absorbance at 600 nm. This is the zero value. Incubate the tubes at the desired temperature and measure the absorbance at 5 min intervals, till there is no change. Draw the time course on a graph sheet. Generally the graph is exponential. The rate of change is calculated from the linear portion of the graph. The molar extinction coefficient of pure dye is 19.1×10^3. Thus, from the change in absorbance the amount of dye reduced in mole unit can be calculated.

Reference

Arrigoni, O. and Singer, T.P. (1962), Nature, 193: 1256.

7.9 DETERMINATION OF PYRUVATE KINASE (EC 2.7.1.40) ACTIVITY

Pyruvate kinase catalyzes the following reaction

$$\text{Phospho(enol)pyruvate} + \text{ADP} \xrightarrow[\text{Mg}^{2+}]{\text{Pyruvate Kinase}} \text{Pyruvate} + \text{ATP}$$

$$\text{Pyruvate} + \beta\text{-NADH} \xrightarrow{\text{Lactic Dehydrogenase}} \text{Lactate} + \beta\text{-NAD}$$

Reagents

1. Potassium Phosphate Buffer 100 mM, pH 7.6 at 37°C.
2. Phospho(enol)pyruvate solution (PEP), 8.0 mM
3. β-Nicotinamide Adenine Dinucleotide, Reduced Form, 3 mM (β-NADH), in phosphate buffer
4. Magnesium sulphate solution, 100 mM.
5. Adenosine diphosphate solution (ADP), 40 mM.
6. L-Lactic Dehydrogenase Solution (LDH)
7. Fructose 1,6-bisphosphate solution (F 1,6-P), 30 mM.

Procedure

Pipette out 1.3 ml of water, 0.8 ml of buffer, 0.16 ml of phosphoenol pyruvate, 0.2 ml of magnesium sulphate, 0.10 ml of ADP, 0.4 ml of LDH and 0.10 ml of fructose 1,6 bisphosphate into a tube labeled test, while blank contains all reagents in the same volume except buffer which is 0.9 ml.

Mix both the tubes by inversion and equilibrate to 37°C, now add 0.2 ml of β NADH and monitor the A_{340}nm until constant, using a suitably thermostatted spectrophotometer. Then add 0.10 ml the enzyme solution to the tube labeled test. Immediately mix by inversion and record the decrease in A_{340}nm for approximately

5 minutes. Obtain the A_{340nm}/minute using the maximum linear rate for both the Test and Blank.

Calculations

$$\text{Units/ml enzyme} = \frac{(A_{340nm}/\text{min Test} - A_{340nm}/\text{min Blank}) \times 3 \times df}{(6.22)\ (0.1)}$$

3 = Total volume (in milliliters) of assay

df = Dilution factor

6.22 = Millimolar extinction coefficient of β-NADH at 340 nm

0.1 = Volume (in milliliters) of enzyme used units/ml enzyme

Unit Definition

One unit will convert 1.0 μmole of phospho(enol)pyruvate to pyruvate per minute at pH 7.6 at 37°C in the presence of 1.0 mM fructose 1,6 diphosphate.

Reference

Cardenas, J.M. and Dyson, R.D. (1973), Journal of Biological Chemistry, 248: 6938-6944.

7.10 DETERMINATION OF (EC 2.7.3.2) ACTIVITY

Creatine kinase catalyses

Adenosine triphosphate + creatine Adenosine diphosphate + creatine → phosphate

Reagents

1. Glycine buffer, 0.1 M, pH 9.0. Add 11 ml of 0.1 N sodium hydroxide to 89 ml 0.1 M glycine-sodium chloride (7.505 g glycine and 5.85 g sodium chloride per litre of solution)
2. Buffered substrate: 0.1 M glycine buffer containing 0.063 M creatine (8.253 g per litre)
3. NADH, 0.001 M adenosine-5-triphosphate, 0.006 M phosphoenol pyruvate, 0.002 M in 2 M glycine buffer ($NADH_2$-ATP-PEP).
4. Lactate dehydrogenase and pyruvate kinase, 2 mg enzyme protein to each per ml., 0.5 M magnesium chloride (LDH-PK).

Procedure

Into each of two tubes place 1 ml fresh sample, 0.7 ml of the $NADH_2$-ATP-PEP reagent and 0.05 ml of the LDH-PK mixture. Mix and place the tubes in a water bath for 10 min at 25°C. Then to one test add 1.75 ml buffered substrate and to the other 1.75 ml of glycine buffer. Mix and transfer to a cuvette. Set the spectrophotometer so that the blank has an extinction of 0.300 at 340 nm and read the test against this. After 10 min again readjust the extinction of the blank to 0.300 and read the test against it.

The unit of activity is that which produces a change in extinction of 0.001 per min.

Reference

Tanzer, M.L. and Gilvarg, C. (1954), J. Biol. Chem. 234: 2301.

7.11 ASPARTATE TRASAMINASE (SGOT) L-ASPARTATE – 2-OXOGLUTARATE AMINOTRANSFERASE (EC 2.6.1.1)

This enzyme catalyses the reaction

L-Aspartate + 2-oxoglutarate → oxaloacetate + L-glutamate

Reagents

1. Buffered substrate: Dissolve 15 g dipotassium hydrogen phosphate, 2 g potassium dihydrogen phosphate and 300 mg 2-oxoglutarate in 800 ml distilled water and add 13.2 g L-aspartatic acid is added. Adjust the pH of the solution to 7.4 with sodium hydroxide and make up to one litre with water.
2. 2,4 – dinitropheny hydrazine (DNPH)- 1mM (200 mg)/litre in 1mol/litre HCl.
3. Sodium hydroxide – 0.4 N
4. Pyruvate solution – 2mM/litre

Procedure

Add 0.2 ml of sample to 1 ml of buffered substrate, mix well and incubate for 60 min in a water bath. Add 1 ml of DNPH and allow to stand for 20 min at room temp and add 10 ml of 0.4N sodium hydroxide. Read the absorbance of the colour developed at 530 nm. Blank contains 1 ml of buffered substrate and 1 ml of DNPH solution with 10 ml of 0.4N sodium hydroxide solution. Graded volumes of standard pyruvate solution are processed similarly. The enzyme activity is expressed as IU/L.

Reference

Mohun, A.F. and Cook. I.J.Y. (1957), J. Clin. Path. 10: 394.

7.12 ALANINE TRASAMINASE (SGPT) (L-ALANINE 2 OXOGLUTARATE AMINO TRANSFERASE, EC. 2.6.1.2)

The reaction catalyzed by the enzyme is
L-alanine + 2-oxoglutarate → pyruvate + L-glutamate

Reagents

1. Buffered substrate: Dissolve 15 g dipotassium hydrogen phosphate, 2 g potassium dihydrogen phosphate and 300 mg 2-oxoglutarate in 800 ml distilled water and add 17.8 g L-alanine. Adjust the pH of the solution to 7.4 with sodium hydroxide and make up to one litre with water.
2. 2,4 – dinitropheny hydrazine (DNPH)- 1mM (200 mg/litre) in 1mol/litre HCl.
3. Sodium hydroxide – 0.4 N
4. Pyruvate solution – 2 mM/litre

Procedure

Add 0.2 ml of sample (serum) to 1 ml of buffered substrate, mix well and incubate for 30 min at 37°C. Add 1 ml of DNPH and allow to stand for 20 min at room temp. Then add 1 ml of 0.4N sodium hydroxide and the colour developed is read at 530 nm. Blank contains 1 ml of buffered substrate and 1 ml of DNPH solution with 10 ml of 0.4N sodium hydroxide solution. A standard curve is prepared using different dilutions of standard pyruvate solution.

The enzyme activity is expressed as IU/L.

Reference

Mohun, A.F. and Cook. I.J.Y. (1957), J. Clin. Path. 10: 394.

7.13 ALKALINE PHOSPHATASE (EC 3.1.3.1)

Phosphatases are enzymes which hydrolyse organic phosphate esters liberating an alcohol derivative of the substrate and inorganic phosphorus (Pi). Alkaline phosphatase gives maximum activity at alkaline pH (8–10). For assaying phophatases ρ-nitrophenyl phosphate is used as the substrate which is hydrolyzed to ρ nitrophenol and Pi. ρ nitrophenol is coourless at acidic or neutral pH but at alkaline pH of 11 gives yellow colour.

Reagents

1. Disodium phenyl phosphate 0.01 M. Dissolve 1.09 g in water and make up to 500 ml. Bring quickly to the boil, cool and add a little chloroform and store in a refrigerator.

2. Sodium carbonate-sodium bicarbonate buffer, 0.1M. Dissolve 3.18 g of anhydrous sodium carbonate and 1.68 g of sodium bicarbonate in water and make up to 500 ml.
3. Buffered substrate. Mix equal volumes of 1 and 2. The pH should be 10.
4. Phenol reagent of Folin and Ciocalteau.
5. Sodium carbonate – 20%
6. Standard phenol solution: stock solution- 100 mg of phenol per 100 ml of solution. Working standard is prepared by diluting stock (1 in 10).

Procedure

Pipette 6 ml of the buffer substrate into a test tube and place in a water bath at 37°C for a few minutes. Add 0.3 ml of sample, mix and allow to remain in the bath exactly for 15 minutes. Then remove and immediately add 2.7 ml of the diluted phenol reagent. At the same time set up a tube for the control containing 6 ml of substrate and 0.3 ml of sample, to which is added immediately 2.7 ml of diluted phenol reagent. Mix well and centrifuge.

Take 4 ml of supernatant and add 1 ml of 20% sodium carbonate. Set up a standard prepared by adding 1 ml of reagent. Place the tubes in a water bath at 37°C for 15 minutes and read the colour at 680 nm against a blank which contains 2.8 ml of water and 1.2 ml of diluted phenol reagent and 1 ml of 20% sodium carbonate.

Activity is expressed as IU/L or King and Armstrong units.

Reference

King, E.J. and Armstrong, A.R. (1934), Canad. Med. Assoc. J., 31: 376.

7.14 ACID PHOSPHATASE (EC 3.1.3.2)

The King-Armstrong method alkaline phosphatase was adapted for the estimation of acid phosphatase by substituting the buffer so that the reaction could be carried out pH 4.9.

Reagents

1. Disodium phenyl phosphate, 0.01M. Dissolve 1.09 g in water and make up the volume to 500 ml.
2. Citric acid – sodium citrate buffer: Dissolve 21 g of crystalline citric acid in water and add 188 ml of 1N sodium hydroxide and make upto 500 ml with water. Adjust the pH to 4.9.
3. Buffer substrate: Prepare fresh by mixing equal quantities of the above 2 solutions.

Procedure

The method is almost the same as that given for alkaline phosphatase. Use 6 ml of the buffer substrate with 0.3 ml of sample and incubate for one hour at 37°C. The procedure and calculation are identical to that of alkaline phosphatase.

Activity is expressed as IU/L or King and Armstrong units.

Reference

King, E.J. and Armstrong, A.R. (1934), Canad. Med. Assoc. J., 31: 376.

7.15 SODIUM POTASSIUM DEPENDENT ADENOSINE TRIPHOSPHATASE (Na^+, K^+-ATPASES) (EC 3.6.3.9)

The activities of phosphatases are determined and measured by estimating the liberated inorganic phosphate from ATP.

Reagents

1. Tris – HCl buffer pH 7.4, 184 mM.
2. Magnesium sulphate – 50 mM
3. Potassium chloride – 50 mM
4. EDTA – 1 mM
5. ATP – 40 mM
6. NaCl – 600 mM
7. TCA – 10%
8. Ammonium molybdate – 2.5% in 3N H_2SO_4
9. ANSA – Dissolve 500mg of 1 amino – 2-napthol-4-sulphonic acid in 195 ml of 15% sodium bisulphate solution and add 5 ml of 20% sodium sulphate solution. Store the solution in brown bottle.
10. Standard phosphate solution: Dissolve 35.1 mg of potassium dihydrogen phosphate in 100 ml of water (conc.80 µg/ml).

Procedure

The reaction mixture contains 1.0 ml buffer and 0.2 ml each of magnesium sulphate, potassium chloride, EDTA and ATP solutions. Equilibriate for 10 min at 37°C in an incubator. Initiate the reaction by adding 0.2 ml of the sample (membrane extract). Incubate the assay medium for 15 min at 37°C. Terminate the reaction and estimate the liberated phosphorus. Make up the volume of the supernatant along with aliquot of standards to 5 ml with water. Add 1ml of ammonium molybdate and 0.5 ml of ANSA and mix well. The blue colour developed is read at 620 nm after 20 min against a reagent blank. The enzyme activity is expressed as µmoles of phosphorus liberated/hr/mg protein.

Reference

Daemen, F.J., de Pont J.J., Lion, F. and Bonting, S.L. (1970), Vision Res. 10: 435.

7.16 CALCIUM DEPENDENT ADENOSINE TRIPHOSPHATASE (Ca^{++} ATPASE) (EC 3.6.1.3)

Reagents

1. Tris – HCl buffer pH 8.0, 125 mM
2. Calcium chloride – 50 mM
3. ATP – 10 mM

Procedure

Mix 0.1ml of Tris-HCl buffer, 0.1 ml each of $CaCl_2$, ATP and water together with 0.1 ml of enzyme to give a total volume of 0.5 ml. Incubate the mixture at 37°C for 15 min and arrest the reaction by adding 0.5ml of ice-cold 10% TCA. The amount of phosphorus liberated is estimated by the method of Fiske and Subbarow.

The enzyme activity is expressed as μ moles of phosphorus liberated/hr/mg protein.

Reference

Hjerten, S. and Pan, H. (1983), Biochim. Biophys. Acta. 728: 281.

7.17 MAGNESIUM DEPENDENT ADENOSINE TRIPHOSPHATASE (Mg^{++} ATPASE) (EC 3.6.1.4)

Reagents

1. Tris – HCl buffer pH 8.0, 125 mM
2. Magnesium chloride – 50 mM
3. ATP – 10 mM

Procedure

Mix 0.1ml of Tris-HCl buffer, 0.1 ml each of $MgCl_2$, ATP and water together with 0.1 ml of enzyme to give a total volume of 0.5 ml. Incubate the mixture at 37°C for 15 min and arrest the reaction by adding 0.5 ml of ice-cold 10% TCA. The amount of phosphorus liberated is estimated by the method of Fiske and Subbarow.

The enzyme activity is expressed as μ moles of phosphorus liberated/hr/mg protein.

Reference

Ohnishi, T., Suzuki, Y. and Ozawa, K. (1982). Biochim. Biophys. Acta. 684:67.

7.18 5' NUCLEOTIDASE (EC 3.1.3.5)

The enzyme catalyzes the reaction

$$5' - \text{AMP} + H_2O \longrightarrow \text{Adenosine} + P_i$$

Reagents

1. Tris – HCl buffer, pH 7.5 – 184 mM
2. Magnesium sulphate – 50 mM
3. Potassium chloride – 650 mM
4. EDTA – 1mM
5. 5' AMP

Procedure

Mix 1.0 ml of Tris-HCl buffer and 0.1 ml each of magnesium sulphate, KCl, EDTA, 5' AMP and water. Initiate the reaction by adding 0.2 ml of enzyme and incubate at 37°C for 15 min. Terminate the reaction by adding 2.0 ml of TCA (10%) and estimate the liberated phosphorus.

The activity is expressed as µmole of phosphorus liberated / mg protein / hour.

Reference

Ito, R. and Tsushima, K. (1972), 17: 906.

7.19 CHOLINESTERASE (EC 3.1.1.8)

Cholinesterase hydrolyzes esters of choline to give choline and acetic acid.

$$CH_3CO.O.CH_2CH_2.N(CH_3)_3 + H_2O \longrightarrow CH_3COOH + CH_2OH.CH_2N(CH_3)_3$$

Reagents

1. *Buffer solution:* Dissolve 10.3 g of sodium barbitone in about 300 ml of water and slowly add 60 ml of normal hydrochloric acid. Add 5.3 g of anhydrous sodium carbonate, warm gently with stirring. Cool to room temperature and make up the volume to 500 ml.
2. *Salt mixture:* Dissolve 4.2 go anhydrous magnesium chloride and 0.2 g of potassium chloride in water and make up the volume to 100 ml.
3. *Acetylcholine bromide:* 0.5 M solution.

4. *Acetylcholine bromide buffer-salt mixture:* Immediately before use mix 8 volumes of the buffer solution and 1 volume each of acetylcholine bromide solution and the salt mixture.
5. Hydroxylamine hydrochloride, 14%.
6. Sodium hydroxide – 14%.
7. Alkaline hydroxylamine: Mix equal volumes of solution (5) and (6).
8. Ferric chloride solution: Dissolve 10 g of ferric chloride in 1 litre of 0.02 N hydrochloric acid.
9. Hydrochloric acid – 0.5 N
10. Acetylcholine bromide standard: Dilute 0.5M solution 1 in 10. This contains 50 μmoles/ml.

Procedure

The test contains 0.2 ml of the sample while 0.2 ml of water is taken as blank. To each tube add 2 ml of acetylcholine bromide-buffer-salt mixture previously warmed to 37°C. Incubate for one hour at 37°C. Then add 2 ml of alkaline hydroxylamine solution followed by 6 ml of 0.5 N hydrochloric acid after 1 min. Stopper and invert the tubes thrice. Pipette 0.5 ml of each into separate tubes and add 10 ml ferric chloride solution. Shake the tubes by inverting and centrifuge. Read the absorbance of the supernatant at 540 nm using a reagent blank. A standard curve is set using acetylcholine bromide solution (20–100 μmoles).

Activity is expressed as IU/L.

One unit of the enzyme corresponds to hydrolysis of one μmole of acetylcholine per ml.

Reference

de la Huerga, J., Yesinick, C. and Popper, H. (1952), Amer. J. Clin. Path. 22: 1126.

Lipid Peroxidation and Antioxidants

8.1 LIPID PEROXIDATION

Lipid peroxide level in plasma is determined using thiobarbituric acid (TBA) reaction. In this method the elimination of water soluble substances that react with TBA is attained by precipitating lipid peroxides along with plasma proteins in a phosphotungstic acid system. The interference from sialic acid and bilirubin both of which react with TBA, is avoided by carrying out TBA reaction in an acetic acid solution.

Reagents

1. Sulphuric acid N/12
2. 5% phosphotungstic acid
3. TBA reagent: Mix equal volumes of 0.67% TBA aqueous solution and glacial acetic acid.
4. n-Butanol
5. Standard malondialdehyde (MDA): Prepare stock standard solution of MDA in distilled water using 1,1,3,3,-tetraethoxypropane. Store at 4°C and dilute just before use, to working standard of 50 nmoles/ml.

Procedure

To 0.02 ml of plasma, add 0.4 ml of N/12 sulphuric acid mix and then add 0.5 ml of 10% PTA. Centrifuge the contents and discard the supernatant. To the sediment add 0.2 ml of N/12 H_2SO_4 and 0.3 ml of PTA. Centrifuge and dissolve the precipitate in 4.0 ml of distilled water. To this add 1.0 ml of TBA reagent and the heat the in a boiling water bath for 60 min. Cool, add 5 ml of n-butanol and vortex the contents. Centrifuge for 20 min and read the absorbance of the supernatant at 515 nm.

To standard malondialdehyde in the range of 2-10 nmoles add 1.0 ml TBA reagent and process in the same manner.

MDA values are expressed as nmoles /dl plasma.

Reference

Yagi, K. (1972), Biochem. Med. 15: 212.

8.2 CONJUGATED DIENES

The extent of peroxidation is determined by the measurement of conjugated diene content, which is arrived at by computing the ratio of absorbance at 233 nm and 215 nm.

Procedure

Evaporate the aliquots of lipids extracts to dryness. Suspend the lipid residues in methanol and measure the absorbance at 215 and 233 nm against a solvent blank.

Conjugated diene content is expressed as the ration of A_{233}/A_{215}

Reference

Recknagel, R.O. and Glende, E.A. (1984), Methods in Enzymology 582: 331.

8.3 CATALASE

This enzyme catalyses the reaction

$$2H_2O_2 \longrightarrow 2H_2O + O_2$$

and the oxygen liberated is estimated.

Reagents

1. Dichromate: acetic acid reagent –5% potassium dichromate in glacial acetic acid
2, Hydrogen peroxide – 0.2M
3. Phosphate buffer – pH 7.0, 0.01M

Procedure

To 0.1 ml of sample add 1.0 ml buffer and 0.4 ml of water. Initiate the reaction by adding 0.5 ml of H_2O_2 and incubate the reaction mixture at 37°C for one min.

Terminate the reaction by adding 2.0 ml of dichromate – acetic acid reagent. Treat standard H_2O_2 solution in the range 4-20 μmoles in the same manner. Heat the tubes in a boiling water-bath for 15 min, cool and measure the absorbance at 570 nm.

Catalase activity is expressed as μmoles of H_2O_2 liberated /min/mg protein.

Reference

Sinha, A.K. (1972), Anal. Biochem. 47: 389.

8.4 SUPEROXIDE DISMUTASE

Superoxide dismutase catalyses the dismutation of superoxide anions (O_2^-) to H_2O_2 and molecular oxygen in the following manner:

$$H_2O_2 + 2O_2^- + 2H + \longrightarrow 2\,H_2O_2 + O_2^-$$

Reagent

1. Carbonate – bicarbonate buffer, pH 10.2, 0.1 M containing 57 mg/dl EDTA.
2. Epinephrine – 3 mM
3. Chloroform AR grade
4. Ethanol AR grade

Procedure

Dilute 0.5 ml of sample (plasma, hemolysate etc) with 0.5 ml of water. Add 0.25 ml of ethanol and 0.15 ml of chloroform (all reagents chilled) and centrifuge. Use the supernatant for the assay.

To 0.5 ml of supernatant add 1.5 ml of buffer. Initiate the reaction by the addition of 0.4 ml of epinephrine and measure changes in optical density per minute at 480 nm. Change in optical density per minute measured as 50 % inhibition of epinephrine to adrenochrome transition by the enzyme is taken as one enzyme unit.

The enzyme is expressed as units/mg Hb (or) protein.

Reference

Misra, H.P. and Fridovich, I. (1972), J. Bio. Chem. 247: 3170.

8.5 GLUTATHIONE PEROXIDASE

Glutathione peroxidase catalyses the reaction

$$2GSH + 2\,H_2O \longrightarrow GSSG + H_2O$$

Reagents

1. Phosphate buffer pH 7.0, 0.32 M
2. EDTA – 0.8 mM
3. Sodium azide – 10 mM
4. Glutathione reduced – 4 mM
5. Hydrogen peroxide – 2.5 mM
6. TCA – 10%
7. Disodium hydrogen phosphate – 0.3 M
8. DTNB – 40 mg of DTNB in 1% sodium citrate
9. Standard glutathione (reduced) containing a concentration of 10 mM was prepared.

Procedure

Mix 0.4 ml of buffer, 0.2 ml each of EDTA, sodium azide, glutathione and hydrogen peroxide and preincubate at 37°C for 10 min. Then add 0.1ml of the sample and incubate at 37°C for 10 min. Terminate the reaction by adding 0.5 ml of TCA. Centrifuge the reaction mixture and to 0.5 ml of the supernatant add 3 ml of disodium hydrogen phosphate and 1 ml of DTNB. Measure the absorbance of the colour developed immediately at 412 nm. Take suitable aliquots of standard solution and process similarly.

Enzyme activity is expressed as µg GSH utilized/min.

Reference

Rotruck, J.T., Pope, A.C., Ganther, H.E., Swanson, A.B., Hafeman, D.G. and Hoekstra, N.G. (1973), Science, 179: 588.

8.6 GLUTATHIONE REDUCTASE UNITS

Glutathione reductase catalyzes the reduction of oxidized glutathione (GSSG) by NADPH or NADH to reduced glutathione (GSH).

$$\text{NADPH (NADH)} + \text{H}^+ + \text{GSSG} \longrightarrow \text{NADP (NAD}^+\text{)} + 2\text{GSH}$$

The activity of the enzyme is measured by following the oxidation of NADPH spectrophotometrically at 340 nm.

Reagent

1. Phosphate buffer, 0.3 M, pH 6.8
2. EDTA – 25 mM
3. GSSG – 1.25 mM
4. NADPH – 3 mM

Procedure

Mix 1.5 ml buffer, 0.5 ml EDTA and 0.23 ml GSSG and equilibriate this reaction mixture at 37°C for 10 min. Add 0.2 ml of sample and 0.1 ml of NADPH to initiate the reaction. Record the change in absorbance at 340 nm against the blank spectrophotometrically.

Activity of glutathione reductase is expressed as MgGSH liberated/min

Reference

Erden, M. and Bor, N.M. (1984), Biochem. Med. 31: 217.

8.7 GLUTATHIONE S-TRANSFERASE (GST) ACTIVITY

GST catalyzes the reaction

$$\text{Glutathione} + \text{CDNB} \longrightarrow \text{Glutathione} - \text{S} - \text{CDNB}$$

Reagents

1. Phosphate buffer – pH 6.5, 0.5 M
2. 1 chloro-2,4 dinitrobenzene (CDNB) – 25 mM in 95% ethanol.
3. Reduced glutathione – 20 mM

Procedure

To 0.4 ml of phosphate buffer add 0.04 ml of chlorodintirobenzene (CDNB) in ethanol. To the blank add 1.46 ml of water and to the test 1.3 ml of water. Incubate the tubes at 37°C for 10 min. After incubation add 0.1 ml of GSH. Mix the contents thoroughly and add 0.1 ml of the suitably diluted sample. Monitor the changes in absorbance at 340 nm for 3 min in a spectrophotometer. Enzyme activities are expressed as IU/L.

Reference

Habig, W.H., Pabst, M.J., Jakoby, W.B. (1974). J. Biol. Chem. 249: 7130.

8.8 GLUCOSE – 6 PHOSPHATE DEHYDROGENASE

Glucose – 6 phosphate dehydrogenase catalyzes the oxidation of glucose - 6 phosphate to 6 phosphoglucanolactone.

$$\text{Glucose 6 phosphate} + \text{NADP} \longrightarrow \text{6 phosphoglucanolactone} + \text{NADPH} + \text{H}^+$$

The assay of glucose 6 phosphate dehydrogenase activity measures the rate of reduction of $NADP^+$ to NADPH.

Reagent

1. 0.05 M Tris – HCl buffer, pH 7.5
2. 1M magnesium chloride
3. 0.01 M NADP solution
4. 0.005% phenazine methosulphate solution in water. Store at 4°C and protected from light.
5. 0.01% 2,6 dichlorophenol indophenol dye in water.
6. 0.02 M glucose – 6 phosphate in water.

Procedure

Mix 1.0 ml Tris – HCl buffer, 0.1 ml of $MgCl_2$, 0.1 ml of NADP, 0.5 ml of phenozine methosulphate, 0.4 ml of 2,6 dichlorophenol indophenols dye, 1.9 ml water and 1.0 ml enzyme solution to make a total volume of 5.0 ml. Allow the mixture to stand at room temperature for 10 min to permit the oxidation of endogenous materials.

Initiate the reaction by adding 0.5 ml of glucose-6 phosphate solution. Measure the absorbance at 640 nm against a water blank at one minute intervals for 3 – 5 min. The activity is arrived from the formula:

$$\text{Activity} = \frac{\text{change in OD}}{\text{minute coefficient}} \times \text{Molar extinction} \times \text{Tf}$$

Temperature coefficient factor Tf at 30°C of 0.76 and molar extinction coefficient of NADPH is 6/17.6. The activity is expressed in terms of milli international units/mg protein.

Reference

Glock, G.E. and Mc Lean, P. (1953), Biochem. J. 55:400.

8.9 γ-GLUTAMYL TRANSFERASE

γ-glutamyl transferase catalyses the reversible transfer of γ-glutamyl groups from γ-glutamyl donors to amino acids or peptides or to water.

Reagents

1. Substrate – 30.3 mg of L-gamma glutamyl p-nitroanilide in 10 ml. The substrate is sparingly soluble. Dissolve by warming to 50 – 60°C. The substrate solution is used within 2 hrs of its preparation.
2. Tris-HCl buffer pH 8.2, 0.1 M
3. Dissolve 13.2 mg of glycine glycine in 10ml of water. This is the second substrate.

4. Acetic acid – 10%
5. Standard p-nitroanilide – Dissolve 13.8 mg p-nitroaniline (recrystalized) in 100 ml of distilled water.

Procedure

Add 0.5 ml of enzyme to the incubation mixture containing 0.5 ml L-gamma glutamyl p-nitroanilide, 2.2 ml of glycineglycine and 1.0 ml buffer. After incubation for 30 min at 37°C arrest the reaction by the addition of 1.0 ml of 10% acetic acid.

Measure the amount of p-nitroaniline liberated in the supernatant, as the difference in optical density at 410 nm between samples, with and without substrate. The substrate incubated in the absence of enzyme under the same conditions is used as a reference blank. Absorbance of p-nitroaniline in the range 15 - 75μg serves as standard curve for arriving at the amount of product formed.

Enzyme activity is expressed as Units/mg protein where one unit is defined as the amount of enzyme required/μmole p-nitoraniline/min under incubation conditions.

Reference

Khedouri, E. and Meister, A. (1965), J. Biol. Chem. 240:3357.

8.10 TOTAL THIOLS

The method of estimation of thiols is based on the development of a yellow colour when DTNB is added to compounds containing -SH groups to form 2 nitro 5-mercapto benzoic acid.

Reagents

1. Sodium phosphate buffer pH 6.8, 0.5 M containing 1g/L of ascorbic acid.
2. DTNB reagent – 150 mg DTNB in 100 ml of 0.5 M sodium phosphate buffer, pH 6.8.
3. Sulphosalicylic acid – 5%
4. Standard glutathione – 100 μg/ml was prepared in 5% sulphosalicylic acid.

Procedure

To 1.0 ml of sample add 4 ml of ice-cold 5% sulphosalicylic acid and centrifuge. To 1 ml of the supernatant add 5ml of 0.5 M phosphate buffer (containing L-ascorbic acid), followed by 1.0 ml of DTNB reagent. Treat standard glutathione (10-50 μg) in the similar manner. Blank containes 1.0 ml of 5% sulphosalicylic acid. The absorbance is measured at 412 nm.

Total thiol content is expressed as μg glutathione /mg protein.

Reference

Ball 1966.

8.11 REDUCED GLUTATHIONE

Reagent

1. Precipitating reagent – Dissolve 1.67g of H_3PO_4, 0.12g EDTA disodium salt, 30 g NaCl in water and make up the volume to 1 litre. This is kept at 4°C.
2. Disodium hydrogen phosphate (DSHP)–0.3 M
3. Disodium salt of EDTA – 0.1%
4. 5-5'-dithiobis-2-nitrobenzioc acid reagent (DTNB) – Dissolve 40 mg of DTNB in 100 ml of 1% sodium citrate. This is stable at 4°C for over 3 months.
5. Standard solution – 10 mg of reduced glutathione in 100 ml of water.

Procedure

Take 0.2 ml of the sample add 3 ml of precipitating reagent and mix thoroughly. Allow to stand for 5 min before filtering; add 2.0 ml of filtrate to 4.0 ml of 0.3M DSHP solution and 1.0 ml of DTNB reagent. Prepare a blank in a similar manner using distilled water instead of sample. A set of standards containing 20 – 70 μg of reduced glutathione are taken and made up to 2.0 ml with water. To each tube add 4 ml of disodium hydrogen phosphate and 1.0 ml of DTNB reagent. The optical density is read at 412 nm.

The values are expressed as μmoles/l blood.

Reference

Beutlar, E. and Kelly, B.M. (1963), Experentia, 19: 96.

8.12 TOTAL SULPHYDRYL GROUPS (TSH)

Reagents

1. Tris HCl buffer, 0.2 M, pH 8.2, containing 0.02M EDTA
2. DTNB – Dissolve 99 mg of dinitro benzene in 2.5 ml of absolute methanol.
3. Chloroform –Isopropanol reagent 3:2 v/v.
4. Standard glutathione – Dissolve 10 mg reduced glutathione in 100 ml water.

Procedure

To 2.9 ml of buffer solution add 0.02 ml of sample and 0.1 ml DTNB and mix well, to this add 3 ml of the chloroform : isopropanol reagent and vortex thrice at 5 min interval. Centrifuge the reaction mixture at 3000 xg at room temperature for 10 min measure the absorbance of clear supernatant at 412 nm. Process standards containing 20 – 100 μg GSH similarly.

Total thiol group is expressed in terms of μg /ml.

8.13 NON PROTEIN BOUND SULPHYDRYL GROUPS (NP-SH)

Reagents

1. Tris HCl buffer, 0.4M, pH 8.9
2. DTNB – Dissolve 99 mg of dinitro benzene in 25 ml of absolute methanol.
3. Standard glutathione – Dissolve 10 mg of reduced glutathione in 100 ml water.

Procedure

Mix 0.2 ml of the sample with 4 ml water and 1 ml TCA and vortex for 10-15 min and centrifuged at 3000 xg for 15 min. Mix 2 ml supernatant with 4 ml of Tris buffer and 0.1 ml of DTNB reagent. The absorbance of the colour developed is read within 5 min of addition of DTNB reagent at 412 nm against a reagent blank of 2.0 ml water. Process standards containing 20-100 μg similarly.

Non–protein sulphydryl groups are expressed as μg/ml.

Protein bound–SH is calculated by subtracting non-protein SH values from Total – SH groups.

Protein bound–SH groups are also expressed as μg/ml.

Reference

Seedlack, J. and Lindsay, R.H. (1968), Anal. Biochem, 25: 192.

8.14 CERULOPLASMIN

Ceruloplasmin is an oxidase and its measurement is based on oxidation of paraphenylenediamine to form a pink coloured complex.

Reagents

1. Acetate buffer – pH 5.5, 1.0 M
2. Sodium azide – 0.5%
3. Paraphenylenediamine dihydrochloride – 0.5%

Procedure

Pipette out 0.1 ml of sample into two 15 ml test tubes, one each for test and control. To the control tube add 1 ml of 0.5% sodium azide. Then add 8 ml of acetic acid buffer to both the tubes followed by paraphenylenediamine dihydrochloride and vortex the contents. Incubate at 37 ºC for 1 hr and add 1 ml of sodium azide to the test. Mix the contents well and cool the tubes at 4 – 10°C for 30 min. Measure the absorbance of colour developed at 530 nm with the control tubes as blank. Serum ceruloplasmin values are expressed as mg/l of serum by multiplying by the factor of 875.

Reference

Henry, R.J., Chiamori, N., Jacobs, S.L. and Seaglove, M. (1960), Proc. Soc. Exp. Biol. Med., 104: 620.

8.15 RETINOL AND CAROTENES IN SERUM

Proteins are precipitated with ethanol and the retinol and carotenes extracted into light petroleum. After reading the intensity of the yellow colour due to the carotenes the light petroleum is evaporated off and the residue dissolved in chloroform. Carr – Price reagent is added and the amount of blue colour produced is read. Since carotenes also give some colour, a correction for this is made in order to obtain that due to the retinol present.

Reagent

1. Absolute ethanol
2. Light petroleum
3. Chloroform
4. Acetic anhydride
5. Carr – Price reagent – Antimony trichloride 250 g/l in chloroform. Keep at room temp in a tightly stoppered brown bottle, filtering before use if necessary.
6. Standard solution – 500 mg of β carotene/l in light petroleum. Working standard contained 10 mg/l.

Procedure

Pipette 3 ml serum into a stoppered centrifuge tube and add 3 ml absolute ethanol slowly drop by drop with shaking in order to obtain a finely divided precipitate of protein. Add 6 ml light petroleum and shake vigorously for 10 min then centrifuge at low speed for about 1 min. Pipette off as much as possible of the light petroleum layer, taking care not to remove any of the water layer with it.

8.16 ESTIMATION OF CAROTENES

Place the light petroleum ether extract in the colorimeter cuvette and measure the absorbance at 440 nm or with a violet filter using light petroleum as blank. Serum carotene concentration is read directly from this curve. The fact that 2 ml of light petroleum contain the carotenes from 1 ml serum is taken into account.

Note: *Carotenes do not keep well since they easily oxidase.*

Reference

1. *Knible, M.S. (1938), J. Lab. Clin. Med. 24: 1055.*
2. *Kaser, M. and Sketol, J.A. (1943), J. Lab. Clin. Med. 26: 904.*

8.17 VITAMIN A

Reagents

1. Ethanol absolute
2. Light petroleum ether AR grade
3. TCA saturated in chloroform: Dissolve 50 g of TCA crystals in alcohol free chloroform (25 ml). This makes up to about 50 ml of reagent which was stable for 4 – 5 hrs.
4. Vitamin A standard: Dissolve 100 mg of vitamin A acetate in 10 ml of chloroform, freshly before use. The solution gave a concentration of 350 μg vitamin A per ml. 1 ml of stock was diluted to 50 ml with chloroform. This solution contained concentration of 7 μg/ml of vitamin A

Procedure

Mix 1.5 ml of the sample with 1.5 ml of ethanol in a glass stoppered tube and shake vigorously. Add 3.0 ml of light petroleum ether and shake thoroughly for about 10 min in a shaker and allow to stand to obtain a clear supernatant. Draw 2.5 ml of the supernatant for the assay of vitamin A.

To 2.0 ml of the supernatant and to standard vitamin A solution taken in the range 150 -350 μg add 2.0 ml of TCA-chloroform reagent. Measure the absorbance at 620 nm within 15 sec. Results are expressed as μmoles/L plasma.

Reference

Bayfield, R.F. and Cole, E.R. Methods in Enzymology, Vol 67, Academic Press, 189.

8.18 VITAMIN C

The ascorbic acid is converted to dehydroascorbic acid and is coupled with dinitrophenylhydrazine in presence of thiourea as a mild reducing agent. Sulphuric acid then converts the dinitrophenylhydrazone into a red compound which is assayed colorimetrically.

Reagents

1. Tris-HCl buffer pH 7.4, 0.1 M
2. TCA – 5%, 10%
3. Dintirophenyl hydrazine – thiourea – copper sulphate (DTC) – Dissolve 0.4 g of thiourea, 0.05 g copper sulphate and 3 g of DNPH and make up to a total volume of 100 ml in 9N H_2SO_4.
4. 65% H_2SO_4 – Add 680 ml of the concentrated acid to 100 ml of distilled water.
5. Standard ascorbic acid – Dissolve 10 mg of ascorbic acid in 100 ml of 5% TCA. This solution had a conc of 10 μg ascorbic acid/ml.

Procedure

Add 2 ml of 10% TCA to 1.0 ml of the sample, shake the tubes well and allow to stand for 5 min and centrifuge at 2,500 xg for 15 min.

To 1 ml of the supernatant obtained from the above add 2 ml of 5% TCA and 0.1 ml of DTC reagent. Vortex the tubes and incubate at 37°C for 3 hrs. Add 0.75 ml of 65% H_2SO_4 and measure the absorbance of the colour developed at 520 nm after 30min. Blank contains 3.0 ml 5% TCA.

Treat standard ascorbic acid solution in the range of 10 – 50 μg in the same way. Values were expressed as μ moles of ascorbic acid/ml of blood.

Reference

Omaye, S.T., Turnbull, J.D. and Sauberlich, H.E. Methods in Enzymology, Vol 62. Academic Press 1979.

8.19 VITAMIN E (α TOCOPHEROL)

Vitamin E is estimated after extraction, sonification and solvent extraction of lipids to remove interfering substances, tocopherol is estimated spectrophotometrically using bathophenanthroline reagent.

8.20 EXTRACTION OF VITAMIN E FROM PLASMA

Reagents

1. Pyrogallol – 2.0% in absolute ethanol.
2. KOH – saturated solution
3. Hexane

Procedure

To 1 ml of plasma add 2.0 ml of 2% solution of pyragallol in purified ethanol and mix thoroughly. Heat the mixture at 70°C for 2 min. Add 0.3 ml of saturated KOH and mix again. Incubate the mixture again at 70°C for 30 min. Mix the contents of the tubes vigorously for 2 min. Cool and add 4 ml hexane centrifuge the tubes at 1,500 xg for 10 min to separate the 2 phases. The hexane extract (upper phase) is used for the assay of vitamin E.

8.21 VITAMIN E ESTIMATION

Reagent

1. Absolute ethanol
2. Hexane (AR)
3. Bathophenanthroline reagent –0.2% solution of 4,7-diphenyl –1,10 –phenanthroline in absolute ethanol.
4. Ferric chloride reagent –0.001M ferric chloride solution in absolute ethanol. The reagent is prepared before use.
5. Orthophosphoric acid reagent – 0.001M in absolute ethanol.
6. Vitamin E standard – 10 mg of vitamin E is dissolved in 100 ml of absolute ethanol. The solution has a concentration of 100 μg/ml.

Procedure

Evaporate 3 ml of hexane extract to dryness under vacuum. Dissolve the residue is carefully in 3.0 ml of absolute ethanol. Add 0.2 ml of 0.2% bathophenanthroline reagent and mix thoroughly. Perform the assay very rapidly from this point. Care should be taken not to expose the tubes to direct light. Add 0.2 ml of ferric chloride reagent and 0.2 ml of orthophosphoric acid reagent and mix well.

Treat standard vitamin E taken in the range of 50 – 100 μg similarly. Measure the absorbance of the pink colour developed at 536 nm against a reagent blank with ethanol.

The vitamin E values were expressed as μmoles/100 of plasma.

Reference

Emmerie, A. and Engel, C. (1938), Rev. Trav. Chim. 57: 1351.

Molecular Biology

9.1 ISOLATION OF NUCLEIC ACIDS FROM BIOLOGICAL SAMPLES

Nucleic acids are present as nucleoprotein complexes in cells. The major problems encountered in isolation of pure and intact DNA or RNA molecules are degradation of high molecular weight nucleic acids by mechanical damage or by hydrolytic enzymes, and contamination with polysaccharides, proteins and other compounds. Methods of isolation of nucleic acids from different sources are given below:

Isolation of DNA form *E.Coli* Cells

The chromosomal DNA of bacteria is isolated by performing lysis of cells with the enzyme lysozyme and finally precipitating the DNA with ethanol.

Reagents

1. TE buffer, 10 mM, pH 8.0, containing 1 mM EDTA.
2. SDS: 10% w/v
3. Sodium chloride: 5 M solution.
3. CTAB/NaCl solution: Dissolve 4.1 g NaCl and 10 g Cetyltrimethylammoniumbromide (CTAB) are dissolved in 100 ml of distilled water.
4. Chloroform: isoamylaclcohol : isopropanol.

Procedure

Bacterial culture is grown until statiophase and centrifuge to form a pellet. Suspend the pellet in 500 µl TE buffer. Add 30 µl of 10% SDS and mix thoroughly and incubate for 1 h at 37°C. To this add 100 µl of 5M NaCl and 80 µl of CTAB/NaCl, mix well and incubate for 10 min at 65°C. To the above add 1 volume of chloroform: isoamylaclcohol: isopropanol mixture and centrifuge for 4-5 min. To the supernatant add 0.6 volume of isopropanol to precipitate the DNA. Centrifuge the mixture and suspend the pellet in TE buffer.

Isolation of DNA from Animal tissue

Reagents

1. Saline citrate solution, pH 7.4
2. Sodium chloride (0.9%): Dissolve 900 mg of sodium chloride in 100 ml of distilled water.
3. Sodium citrate (0.5%): Dissolve 500 mg of sodium citrate in 100 ml of water.

Procedure

Grind 500 mg of tissue in 5 ml of saline citrate solution, transfer the homogenate into a centrifuge tube and make up the volume to 10 ml with saline citrate solution. Centrifuge the tubes at 3000 rpm for 8 min and discard the supernatant. Resuspend the pellet in 5 ml of saline citrate and make up the volume to 10 ml with saline citrate solution. Centrifuge again for 10 min at 3000 rpm and discard the supernatant. Suspend the pellet in 12% NaCl solution and centrifuge at 10,000 rpm for 15 min, transfer the supernatant to another tube and add 2 – 3 volumes of cold ethanol. Gently mix by inversion and spool the precipitated DNA by winding around a clean glass capillary tube.

9.2 ISOLATION OF GENOMIC DNA FROM CAULIFLOWER

This is a rapid method for isolation of DNA from plants. Plant tissue is homogenized and mixed with 5% TCA to remove acid soluble compounds and then extracted with alcohol to remove phospholipids. The lipid free sediment is incubated overnight with warm alkali (1N KOH) which hydrolysis RNA to soluble nucleotides but do not affect the DNA which is already dissolved in TCA. This method has an advantage that DNA and RNA are separated and can be estimated by determining phosphate sugars and base contents.

Reagents

1. Tricholoroacetic acid 5%
2. Ethanol 95%
3. KOH 1N
4. HCl 5N

Procedure

Grind 3 g of plant tissue and add 3 ml of 5% TCA. Centrifuge the extract and discard the supernatant. Extract the pellet thrice with 5 ml of 95% alcohol, centrifuge and discard the supernatant. Resuspend the pellet in 5 ml of 1N KOH and incubate at 37°C overnight. The following day centrifuge the extract and discard the supernatant. Boil the pellet with 0.5 ml of 5N HCl and 2 ml of 5% TCA. Collect the supernatant and from this DNA is precipitated.

9.3 ISOLATION OF PLANT GENOMIC DNA BY CTAB METHOD

CTAB is a cationic detergent, which aids in the lysis of cell membranes and forms complexes with nucleic acids. Sodium chloride aids in the formation of nucleic acid-CTAB complexes. Since plant cells have both cell wall and cell membranes, they must be removed one by one to release the cell contents. So DNA isolation involves two parts, a technique to lyse the cells gently and solubilize the DNA, followed by one of several basic enzymatic or chemical methods to remove contaminating proteins, RNA and other macromolecules.

Reagents

1. CTAB-extraction buffer

   Tris-HCl (pH 8.0) - 100 mM

   Sodium chloride - 1.4 M

   EDTA (pH 8.0) - 20 mM

   CTAB - 2.0 % w/v

   2 – Mercaptoethanol - 0.1% w/v

   (The extraction buffer must be autoclaved without 2-mercaptoethanol).
2. Liquid nitrogen
3. RNase A solution - 10 mg/ml
4. Chloroform:isoamyl alcohol - 24:1
5. Ethanol - 70%
6. TE Buffer

   Tris-HCl (pH 8.0) - 10 mM

   EDTA (pH 8.0) - 1 mM
7. Absolute alcohol
8. NaCl - 3 M
9. Sodium acetate - 3 M

Procedure

Grind 2 g of leaf tissue in liquid nitrogen using a mortar and pestle and transfer the ground material to a 50 ml centrifuge tube. To the ground material, add 15 ml CTAB extraction buffer, preheated to 90°C and suspend thoroughly. Incubate the suspension at 65 °C for 30 min with occasional mixing. Cool the mixture to room temperature. Add equal volumes of chloroform : isoamyl alcohol mixture (24 : 1) and mix well to get an emulsion by inverting the tube several times for 15 min. centrifuge the mixture at 12,000 rpm for 5 min and take out the aqueous layer carefully.

Precipitate the chromosomal DNA by adding 0.7 volume of isopropanol. Mix carefully and keep at room temperature for 15 min. Retrieve the DNA with a glass hook or a pipette. Transfer the DNA to a clean tube and air dry for 15 min. Dissolve the DNA in minimal volume of TE buffer. Incubating at 65°C for 10 min will accelerate the dissolution of DNA and inactivate DNase I. Add 1/100 volume of 10 mg/ml RNase A solution. Mix by inversion and incubate at 37°C for 30 min. Add equal volume of chloroform : isoamyl alcohol (24 :1) mix well and centrifuge at 12,000 rpm for 10 min.

To the aqueous layer add 2 volumes of absolute alcohol and 1/10 the volume 3M sodium acetate. Spin for 5 min at 10,000 rpm. Discard the supernatant. Wash the precipitate with 70% ethanol. Air dry the pellet and dissolve in TE buffer. Once resuspended in TE the nucleic acid solution can be stored at –20°C until use.

9.4 ISOLATION OF DNA FROM BLOOD

Reagents

Reagent A:

Sucrose, 320 mM

$MgCl_2$, 5 mM

Triton X, 1%

Make upto 1litre with distilled water.

Reagent B

Tris-HCl, 400 mM, pH 8

Na-EDTA, 60 mM

NaCl, 150 mM

Make upto 95 ml with distilled water and add 5 ml of 20% SDS.

Reagent C

Tris-HCl, 10 mM

EDTA, 1mM

Make upto 100 ml with distilled water.

Procedure

To one volume of blood sample add 4 volumes of reagent A and mix gently till the solution becomes clear. Centrifuge the tubes at 2500 rpm for 5 min to obtain a pellet, free from RBC. Discard the supernatant which contains the lysed RBC. Disturb the pellet thoroughly and add half the volume (as that of the sample) of reagent B. Mix thoroughly by inverting the tube for 3 min till the solution becomes viscous. Now add reagent C (one-fourth the volume of reagent B) and mix gently for 3-4 min. add equal volumes of (reagent B&C) of phenol and chloroform mixture (1:24). Mix well and centrifuge at 2500 – 3000 rpm for 7-8 min to separate into 3 layers viz, aqueous layer, protein layer and solvent layer.

Transfer the aqueous layer carefully into another centrifuge tube using a broad mouth tip. Add equal volume of chloroform to the supernatant and mix gently for a minute and centrifuge at 2500 rpm for 5 min. Transfer the aqueous phase to a fresh tube. To this add 2 volumes of chilled absolute alcohol to gently precipitate the DNA. Spool out the DNA into a fresh tube and decant the alcohol. Wash the DNA twice with 70% alcohol and centrifuge to remove the alcohol. Ensure that the pellet is dry and alcohol free. Suspend the pellet in TE buffer; incubate it at 55°C for 45 min to enhance the dissolution. The sample is stored at 4°C.

Reference

Harris, S.A. (1996), Molecular analysis of froest tree biodiversity A selection of practical protocols. Univ of Oxford manual: 30.

9.5 ISOLATION OF RNA FROM YEAST

Yeast cells are homogenized with phenol which dissociates proteins from nucleic acids and also denatures proteins. Upon centrifugation DNA remains in the lower phenol phase, RNA is dispersed in the upper aqueous phase with the denatured proteins being at the interphase. RNA is recovered from the aqueous phase and precipitated with ethanol.

Reagents

1. Dried yeast
2. Phenol solution: Dissolve 900 g of phenol in 1 liter water.
3. Potassium acetate: Dissolve 200 g of potassium acetate in 1 liter of water and adjust its pH to 5.0.
4. Absolute ethanol
5. Diethyl ether

Procedure

Suspend 30 g of dried yeast in 120 ml of distilled water and incubate at 37°C for 15 min. Add 160 ml of phenol solution and stir suspension for 30 min at room temperature and then centrifuge it at 3000 xg for 15 min in cold to facilitate the emulsion to separate out into different phases. Remove the upper aqueous phase and add potassium acetate solution to obtain a final concentration of 20 g of potassium acetate per liter of the mixture. Add two volumes of ice cold absolute ethanol to precipitate out the RNA and leave it on ice for 1h. Centrifuge at 2000 x g for 5 min at 4°C. Discard the supernatant and wash the RNA pellet with about 10 ml of cold 70% ethanol, then absolute ethanol and finally diethyl ether. Centrifuge at each washing step at 2000 xg for 5 min ad collect the pellet.

9.6 ISOLATION OF RNA FROM PLANT TISSUE BY SDS-PHENOL METHOD

SDS and phenol dissociate nucleoprotein complex and denature proteins. Further treatment with phenol : chloroform : isoamyl alcohol removes denatured proteins and RNA is recovered by precipitation with ethanol from the aqueous phase while DNA remains confined in the phenol phase.

Reagents

1. **Phenol reagent:** Dissolve 500 g of redistilled phenol, 500 mg of 8-hydroxyquinoline in 150 ml of water and m-cresol 70 ml.
2. **SDS solution:** 1% SDS, 015 M sodium acetate, 5 mM EDTA, 50 mM Tris-HCl (pH 9.0).
3. **Deproteinizing solution:** Mix phenol reagent with an equal volume of chloroform and add isoamyl alcohol, an antifoaming agent so that its concentration is 4% *v/v*.
4. Absolute ethanol
5. **Sodium acetate buffer:** Prepare 0.15 M solution of sodium acetate and adjust pH to 6.0 with acetic acid and then add SDS so that its concentration is 0.5%.

Procedure

Homogenize the plant tissue with SDS solution. The ratio of tissue weight to the SDS solution should be approximately 1:10. Transfer the homogenate to a centrifuge tube and add an equal volume of deproteinizing solution and mix for 5–10 min at room temperature. Centrifuge the tubes at 10,000 xg for 10 min at 4°C. Three layers will be formed, with the upper aqueous phase, taking care not to include the protein interphase and transfer it to a clean centrifuge tube. If the aqueous phase contains significant amount of proteins, re-extract RNA by repeating the above mentioned steps. Add about 2.5 volumes of chilled absolute ethanol to the aqueous phase, mix well and allow it to stand at – 20°C overnight. Collect the precipitated nucleic acid by centrifuging at 10,000 x g for 10 min at 4°C. Discard the supernatant and air dry the pellet. Add a few ml of sodium acetate buffer to dissolve the pellet, reprecipitate RNA from this solution by adding 2.5 volumes of ethanol and keeping it at 20°C overnight. Centrifuge at 10,000 x g for 10 min at 4°C, and discard the supernatant and wash the pellet with 70% ethanol.

9.7 PLASMID DNA ISOLATION BY ALKALINE LYSIS METHOD

The bacterial cells are lysed by lysozyme and SDS at high pH and the lysate is then neutralized. The plasmid DNA undergoes renaturation under these conditions but the chromosomal DNA remains denatured and gets precipitated out in the form of a protein-DNA-SDS complex.

Reagents

1. Alkaline lysis solution I: Prepare 100 ml of this solution containing 50 mM glucose, 25 mM Tris HCl (pH – 8) and 10 mM EDTA (pH 8)
2. Alkaline lysis solution II: Prepare 100 ml of this solution containing 0.2 N NaOH and 1% *w/v* SDS (should be prepared fresh during use).
3. Alkaline lysis solution III: To 100 ml of this solution containing 60 ml of 5 M potassium acetate, 11.5 ml of glacial acetic acid and 28.5 ml of water.
4. Equilibrated phenol
5. Luria broth (LB) medium
6. Chloroform:isoamylalcohol (25:1)
7. Isopropanol
8. Tris EDTA (TE) buffer

Procedure

Inoculate a single colony of plasmid containing bacteria into 5 ml of LB medium containing appropriate antibiotic. Incubate the culture at 37°C for 16-18 h. Centrifuge the culture at 8,500 rpm for 5 min. Discard the supernatant and resuspend the pellet in 250 µl of ice cold solution of alkaline lysis solution I by vigorous vortex mixing.

Add 300 µl of alkaline lysis solution II. Invert the tube gently for 4-5 times and incubate at room temperature for 5 min. Add 300 µl of alkaline lysis solution III. Mix gently by inverting. Incubate the suspension on ice for 10 min, centrifuge at 12,000 rpm for 15 min. Transfer the supernatant to a fresh tube. To this add

0.4 volume of phenol: chloroform: isoamylalcohol (26:25:1) reagent. Mix well. Centrifuge the contents at 12,000 rpm for 12 min. To the aqueous phase add 0.6 volume of isopropanol at room temperature and incubate for half an hour. Centrifuge at 13,000 rpm for 20 min. To the pellet add 70% ethanol and store at 37°C until the ethanol dries. Resuspend the pellet in 20 µl of TE buffer. Incubate the tubes at 50°C for 45 min. vortex. Estimate the concentration of the isolate by running in 0.8% of agarose gel.

Reference

Brinboim, H.C. and Doly, J. (1979), A rapid alkaline extraction procedure for screening recombinant plasmid DNA. Nucleic Acids Res. 7: 1513.

9.8 MAXI PREPARATION OF PLASMID DNA BY ALKALINE LYSIS WITH SDS

Reagents

1. Alkaline lysis solution I: Prepare 100 ml of this solution containing 50 mM glucose, 25 mM Tris HCl (pH – 8) and 10 mM EDTA (pH 8).
2. Alkaline lysis solution II: Prepare 100 ml of this solution containing 0.2 N NaOH and 1% *w/v* SDS (should be prepared fresh during use)
3. Alkaline lysis solution III: Mix 60 ml of 5 M potassium acetate solution, 11.5 ml of glacial acetic acid and 28.5 ml of water.
4. Ethanol
5. Luria broth (LB) medium
6. Chloroform: Isoamylalcohol (25:1)
7. Isopropanol
8. Streptomycin - 10 mg/ml
9. Tris EDTA (TE) buffer

Procedure

Inoculate the culture into 500 ml LB broth containing the appropriate antibiotic and incubate overnight with vigorous shaking. Centrifuge the culture at 4,100 rpm for 15 min. Discard the supernatant and resuspend the pellet in 200 ml of ice-cold Tris-EDTA buffer. Centrifuge the cells at 4,100 rpm for 15 min. Resuspend the pellet in 18 ml of alkaline lysis solution I. To it add 2 ml of freshly prepared lysozyme solution and 40 ml of freshly prepared alkaline lysis solution II. Mix well and incubate at room temperature for 5-10 min. Add 20 ml of ice cold alkaline lysis solution III and mix well. Keep the tubes on ice for 10 min. Centrifuge the bacterial lysate at 10,000 rpm for 30 min. Decant the supernatant and measure the volume. To this add 0.6 volume of isopropanol and mix well. Incubate at room temperature for 10 min. Recover the precipitated nucleic acid by centrifugation at 8000 rpm for 15 min. Discard the supernatant. Rinse the pellet with 70% ethanol twice and drain. Resuspend the pellet in 3 ml of TE buffer. Purify the plasmid DNA by equilibrium centrifugation in CsCl-EtBr gradient.

9.9 AGAROSE GEL ELECTROPHORESIS

Agarose gel electrophoresis is the means of separating DNA by electrophoresis

Reagents

1. 1.5% agarose
2. 1 X Tris-acetate-EDTA (TAE) buffer (Electrophoresis buffer)
3. Ethidium bromide (EtBr)
4. 6 X gel loading buffer
5. DNA sample

Procedure

Clean the edges of the glass plate and seal with tape to form a mold. Keep the mold in a horizontal position. Fill the electrophoresis tank with sufficient amount of TAE buffer and caste the gel. Prepare 1.5% agarose and keep it in Erylenmeyer flask. Heat the slurry until the agarose gets dissolved. Add EtBr to a final concentration. of 7μg (wear gloves while using EtBr). Mix thoroughly by gently swirling. Fix the comb to a position of 0.5-10 mm of the plate. Pour the warm agarose into the mold. Allow the gel to set. Add a small amount of buffer on the top of the gel. Remove the comb slowly. Mount the gel on the electrophoresis tank. Add the buffer to the electrophoresis tank to cover the gel. Mix the DNA sample with the gel loading buffer. Add the sample to the well using a micropipette. Cover the tank with the lid. Apply 50 V of electric field to it. (You can see bubbles generated both at the cathode and anode). Bromophenol blue starts migrating from the well. Once the DNA sample has migrated to a sufficient distance, turn off the electric current. Remove the lid and examine the gel under UV. Document the result.

Reference

Sambrook, J., Fritsch, E.F and Maniatis, T. (1989). Molecular cloning a laboratory manual 2nd Ed Cold Spring Harbor, New York.

9.10 RNA GEL ELECTROPHORESIS

Reagents

1. 10 X Mops buffer

2. 0.2 M Morpholino propane sulfonic acid (Mops)
3. 0.05 M Sodium acetate
4. 0.01 M EDTA
5. Loading buffer
6. Formamide
7. 80% glycerol
8. Bromophenol blue
9. EtBr

Procedure

Clean the gel apparatus with 1% SDS and rinse with water. Add 1g agarose, 10 ml 10 X Mops buffer, and 85 ml water and place it in boiling water bath for 5 min. and allow it to dissolve. Sterilize agarose gel and cool. Add 37% formaldehyde to solution and mix by swirling. Add 20 µl of EtBr and mix by swirling. Position the comb and pour the agarose gel on to the plate. Add 10 to 15 µg of RNA sample to 20 µl of gel loading buffer Heat the sample to 95°C for 2 minutes to denature RNA. Place the gel slab into the electrophoresis tank. Use 1 X Mops buffer as running buffer. Load the gel with 20 µl of RNA sample per well. Run the gel at 20 V for approximately 2-3 hours until bromophenol dye migrates three fourth down the gel. Photograph the gel.

Reference

Sambrook, J., Fritsch, E.F and Maniatis, T. (1989), Molecular Cloning a Laboratory Manual 2nd Ed Cold Spring Harbor, New York.

9.11 QUANTITATION OF DNA

DNA - Diphenylamine Reaction

This is a general reaction given by deoxypentoses. The 2-deoxyribose of DNA in the presence of acid, is converted to ω-hydroxylevulinic aldehyde which reacts with diphenylamine to form a blue coloured complex with absorbance maxima at 600 nm. Compounds such as furfuryl alcohol arabinal, which can be converted into ω-hydroxylevulinic aldehyde will also give this reaction. In DNA, since only deoxyribose of purine nucleotides is released, the value obtained represents half of the total deoxyribose in the sample.

Reagents

1. **Standard DNA solution:** Dissolve calf thymus DNA (100 µg/ml) in 1 N perchloric acid by heating at 70°C for 15 min.
2. **Diphenylamine solution:** Dissolve 1.5 g of diphenylamine in 100 ml of glacial acetic acid and 1.5 ml of sulphuric acid.

3. **1.6% (w/v) acetaldehyde:** Prepare by dissolving 1 ml of ice cold acetaldehyde in 50 ml of distilled water.
4. **Diphenylamine reagent:** Prepare by mixing 0.5 ml of 1.6% of acetaldehyde and 100 ml of diphenylamine solutions. This solution must be prepared fresh.

Procedure

Take 2.0 ml aliquot of the sample in which DNA has to be estimated in a test tube and in another set of test tubes pipette 2.0 ml of standard DNA solution in different concentrations. In one of the test tubes take 2.0 ml of 0.5 N $HClO_4$ as a reagent blank. Add 0.4 ml diphenylamine reagent to all the tubes, mix the contents properly and keep at room temperature in dark for 16 – 18 h or overnight. The absorbance is read at 600 nm. From the absorbance of the sample the amount of DNA is determined.

Reference

Dische, Z., Mikro. (1930), Chemie. 8: 4.

9.12 DETERMINATION OF RNA BY ORCINOL METHOD

This is a general method for the estimation of pentoses. Acid hydrolysis of RNA releases ribose which in the presence of strong acid undergoes dehydration to yield furfural. Orcinol, in the presence of ferric chloride as a catalyst, reacts with furfural producing a green coloured compound with absorbance maxima at 665 nm. DNA gives a limited positive reaction with orcinol test.

Reagents

1. 5% perchloric acid
2. **Standard RNA:** Dissolve yeast RNA (500 μg/ml) in 5% $HClO_4$. Make different dilution to obtain solutions containing 100 – 500 μg RNA/ml with 5% $HClO_4$.
3. **Orcinol reagent:** Dissolve 100 mg of ferric chloride ($FeCl_3.6H_2O$) in 100 ml of conc. HCl and then add 3.5 ml of 6% solution of orcinol prepared in alcohol.

Procedure

Take 2.0 ml solution of each of the dilutions of RNA standard solution, test sample and 2.0 ml of 5% $HClO_4$ as a blank, in different test tubes. Add 3.0 ml of orcinol reagent to all the tubes and mix well. Keep the tubes in a boiling water bath for 20 min. after cooling add 7.0 ml of n-butanol to each tube and measure the absorbance at 665 nm against a reagent blank. The amount of RNA in the sample is determined from the standard curve.

References

1. *Dische, Z. and Schwartz, K. Mikro. Chim. Acta. 2: 13, (1937).*
2. *Grossman, L. and Moldave K ed. Methods in Enzymology. Vol XII. Nucleic acids-Part B. Academic Press. NY, (1968).*

9.13 QUANTITATIVE ESTIMATION OF DNA AND RNA BY SPECTROPHOTOMETRIC METHOD

Both DNA and RNA exhibit strong absorbtion of ultra-violet light due to the presence of conjugated double bonds of the constituent purine and pyrimidine bases. These have a characteristic absorbance maximum at 260 nm which is linearly related to concentration of DNA and RNA in the solution up to an O.D value of 2.0. This spectrophotmetric method can also be employed for estimating purity of DNA and RNA preparations. Proteins are usually the major contaminants in nucleic acid extracts and these have absorption maxima at 280 nm. The ratio of absorbance at 260 and 280 nm, hence provides a rough idea about the extent of contaminants in the preparations. A ratio between 1.8–2.0 is indicative of fairly pure DNA and RNA preparations but values less than 1.8 signify presence of proteins as impurities.

Reagents

1. Saline sodium citrate (SSC) solution: Prepare 0.015 M solution of sodium citrate (pH 7.0) and dissolve NaCl so that its final concentration in solution is 0.15 M.

Procedure

Adjust the wavelength of spectrophotometer to 260 nm and turn on the UV lamp. Set the instrument at zero absorbance with SSC solution. Read the absorbance of the test solution. The calculation of concentration of DNA and RNA in the sample is done as follows:

(a) *For double stranded DNA:* Concentration of DNA in sample solution (µg/ml) = 50 × A260 × Dilution factor

(b) *For RNA:* Concentration of RNA in sample solution (µg/ml) = 40 × A260 × Dilution factor

Reference

Becker, J.M., Caldwell, G.A. and Zacgo, E.A. (1996), Biotechnology a Laboratory Course, Academic Press, New York.

9.14 DETERMINATION OF MELTING TEMPERATURE (TM) AND BASE COMPOSITION OF DNA FROM THERMAL DENATURATION CHARACTERISTICS

When a dilute aqueous solution of double stranded DNA is heated, the two strands get separated due to disruption of H-bonds between the complementary bases. Such DNA is referred to as denatured DNA and the process is called denaturation. DNA denaturation is a reversible process. If the above solution is allowed to cool slowly to room temperature, the complementary strands of DNA reassemble to give a duplex DNA. DNA is now said to be reannealed or renatured and the process is called reannealing or renaturation of DNA.

Upon denaturation, the absorbance of DNA at 260 nm increases by 30–40% due to exposure of bases. This is called hyperchromic effect. Heat treatment at temperatures of upto 80°C causes only a very slight increase in absorbance at 260 nm and this is followed by a sharp increase within a narrow temperature range (80 - 90°C) and then A_{260} becomes constant when DNA is fully denatured. A graph drawn between absorbance at 260 and temperature gives the thermal denaturation curve or melting curve of DNA. The temperature corresponding to the midpoint of the curve is called melting temperature (Tm) of DNA and it denoted the temperature at which 50% of the DNA has undergone denaturation. Tm is largely dependent on the (G+C) content of DNA, higher the (G+C) content, higher will be the Tm. Under standard conditions of ionic strength and pH, (G+C) content is related with Tm in accordance with the following equation:

$$\%\,(G{+}C) = (Tm - 69.3) \times 2.44$$

Reagents

1. Saline sodium citrate (SSC): Prepare 0.015 M sodium citrate solution (pH 7.0) and dissolve NaCl in it so that its final concentration in solution is 0.15M.
2. DNA: dissolve 50 μg of DNA/ml of SC.

Procedure

Warm the spectrophotometer for sufficient time before the test and set the zero at 260 nm with SSC. Measure the absorbance of DNA sample at 25°C. Heat the DNA solution at a rate of 1°C rise/min upto 100°C in different water baths set at temperatures: 25, 35, 50, 70, 75, 80, 85, 90, 95 and 100°C monitoring the temperature using a thermometer. Record the absorbance values. Calculate A_{260} (T°C)/ A_{260} at 25°C for each of the following and plot the absorbance ratio against the temperature. Determine the midpoint of increase in absorbance and by extrapolation find the corresponding temperature which represents Tm for the DNA sample. Calculate the %(G+C) content of the DNA using the following equation:

$$\% (G+C) = (Tm - 69.3) \times 2.44$$

9.15 RESTRICTION ENZYME DIGESTION

Restriction digestion is a process of cutting DNA molecules into smaller fragments with special enzymes called Restriction Endonucleases. Restriction enzymes are bacterial enzymes which cut DNA into defined and reproducible fragments. These special enzymes recognize specific sequences called palindromic sequence in the DNA molecule and cleaves symmetrically in both strands. Restriction digests begin by mixing the DNA and the restriction enzyme. The actual reaction conditions vary from one enzyme to the next and include temperature, NaCl and/or $MgCl_2$ concentration, pH etc. The ingredients are mixed in appropriate concentration, the tube is incubated at the restriction enzyme's optimal temperature for 1 hour or longer. Then finally the digest is run for the appropriate time, the reaction tube is put back on ice to prevent nonspecific degradation of the DNA. Once the restriction digest is completed, agarose gel electrophoresis is performed to separate the digest fragments by size and visualize the fragments and purify for further experiments.

Reagents

1. Restriction enzyme
2. 10 X enzyme buffer
3. DNA samples
4. 6 X loading dye, 0.5 M EDTA

Procedure

Prepare the reaction mixture containing 0.5 µl restriction enzyme (2-3 units), 1.0 µl enzyme buffer (10 X) and 1.0 µl DNA (1 µg). Make up the volume to 10 µl with sterile distilled water. Always add water and buffer into the test tube first, then add the enzyme, add the DNA last and mix by tapping the tube with the finger. Quickspin to remove bubbles (DNA will adhere to bubble surface and becomes inaccessible to the enzyme). Incubate at the recommended temperature for 1 h. Stop the reaction by adding 2.5 µl 6X loading dye mix if the sample is to be loaded directly onto a gel; otherwise stop the reaction by placing it at –20°C or add 0.5 µl of 0.5 M EDTA. After restriction digestion the sample is loaded on an agarose gel and the fragments are visualized.

References

1. *Becker, J.M., Caldwell, G.A. and Zacgo, E.A. (1996), Biotechnology a Laboratory Course, Academic Press, New York.*

2. *Sambrook, J., Fritsch, E.F. and Maniatis, T. (1989), Molecular Cloning a Laboratory Manual 2nd Ed. Cold Spring Harbor, New York.*

9.16 POLYACRYLAMIDE GEL ELECTROPHORESIS (PAGE)

Electrophoresis is widely used to separate and characterize proteins by applying electric current and is rapid and relatively sensitive requiring only micro-weights of protein. Electrophoresis of protein in polyacrylamide gels is carried out in buffer gels (non denaturing) as well as Sodium Dodecyl Sulphate (SDS) containing denaturing gels. SDS is an anionic detergent which binds strongly to and denatures proteins. The number of SDS molecules bound to a polypeptide chain is approximately half the number of amino acid residues in that chain. The protein-SDS complex carries net negative charges, hence moves towards the anode and the separation is based on the size of the protein.

Reagents

1. Stock acrylamide solution

Acrylamide	–	30% w/v
Bisacrylamide	–	0.8 % w/v

   Make upto 100 ml with double distilled water and filter.

2. Separating gel buffer 4X (pH 8.8)

Tris-HCl	–	1.5 M

   Adjust the pH with 2N HCl.

3. Stacking gel buffer 4X (pH 6.8)

Tris-HCl	–	0.5 M

   Dissolve 3.02 g, adjust pH with HCl and then make upt to 50 ml and store at 4°C).

4. Polymerizing agent

Ammonium persulphate	–	10%w/v
TEMED		
Sodium Dodecyl Sulphate	–	10%w/v

5. Electrode buffer 5X (pH 8.2 – 8.4)

Tris-HCl	–	0.25 M
Glycine	–	1.92 M
SDS	–	10.0%
Water	–	500 ml

5. Sample buffer 4X

Tris-HCl	–	0.5 M	1.0 ml
SDS w/v	–	10%	0.8 ml
Glycerol	–	40%	4.0 ml
2-mercaptoethanol	–	20%	2.0 ml
Bromophenol blue	–	0.5%	50 µl

(Store frozen in small aliquots and diluted to 1X concentration just before use)

Standard marker proteins	*Molecular weight (Daltons)*
L-Lactalbumin	14,200
Trypsin inhibitor protein	20,100
Carbonic anhydrase	29,000
Glyceraldehydes-3-phosphate dehydrogenase	36,000
Albumin (egg)	45,000
Albumin (Bovine)	66,000

The above proteins are in single strength sample buffer at a concentration each of 1.0 mg/ml and the wells are loaded with 25–50 µl.

7. Destaining solution

Methanol	–	40%
Acetic acid	–	10%

8. Distilled water – 50 ml

Staining solution

Coomassie brilliant blue R 250 1% in destaining solution. (First dissolve the dye in methanol and then add acetic acid and water. Use fresh preparation every time).

Prepare separating gel solution following the composition as shown below:

Composition for separating gel solution

Solution required	*Percentage of gel*	
	12.5	*10*
Acrylamide	12.5 ml	10.0 ml
Separating gel buffer	6.0 ml	6.0 ml
Water	11.530 ml	13.530 ml
SDS (10%)	0.30 ml	0.30 ml
APS (5 %)	0.150 ml	0.150 ml
TEMED	0.030 ml	0.030 ml

Then the stacking gel is cast over the separating gel using 4% stacking gel mixture prepared as follows for 10 ml (Solution B)

Stock acrylamide	–	1.66 ml
Stacking gel buffer	–	1.38 ml
Distilled water	–	6.86 ml

The mixture is degassed for 2–3 min and then add the following are added

APS	–	100 µl
SDS	–	100 µl
TEMED	–	10 µl

Staining solution

Dissolve 100 mg of Coomassie Brilliant Blue R 250 in 40 ml methanol and 10 ml glacial acetic acid and make up the volume to 100 ml with distilled water.

Destaining solution

Methanol	–	40 ml
Glacical acetic acid	–	10 ml
Distilled water	–	50 ml.

Procedure

Pour the acrylamide solution for separating gel (Solution A) between the plates with appropriate spacer. Mix thoroughly and pour the matrix uniformly without air bubbles. The meniscus of the acrylamide solution should be far below the top of the notched plates to allow for the length of the teeth on the comb plus 1 cm.

Carefully overlay the acrylamide solution with isobutanol or water. For gels made with acrylamide concentration lower than 8% use water; for gels of 10% or greater use isobutanol or water saturated isobutanol. The overlaying solution creates a barrier to oxygen, which inhibits the polymerization of acrylamide. Place the gel in a vertical position at room temperature for polymerization.

After the gel has set, pour off the overlay and wash the top to the separating gel several times with distilled water. Drain well to dry with the edge of a paper towel. Pour the acrylamide solution for stacking gel (Solution B) directly onto the polymerized separating gel.

Place the appropriate comb into the gel solution, being careful not to trap any air bubbles. Place the gel in a vertical position at room temperature. The stacking gel will set in approximately 30 min. Remove the comb without distorting the shape of the wells after the stacking gel has polymerized. Wash the wells immediately with distilled water to remove unpolymerised acrylamide by flushing with a syringe or a pipette. Straighten the teeth of the staking gel, if necessary, and place in an appropriate vertical gel electrophoresis system with running buffer at the bottom reservoir.

Add running buffer to top reservoir. Any bubbles caught between the plates at the bottom of the gel can be removed by squirting running buffer through a syringe. The electrode buffer and the plates can be kept cooled using a suitable facility so that heat generated during the run is dissipated and does not affect the gel and resolution.

Prepare the samples for electrophoresis, following suitable extraction procedure. Adjust the protein concentration in each sample using the five strength sample buffer and water in such a way that the same amount of protein is present per volume. Also the concentration should be such as to give a sufficient amount of protein (50–200 µg) in a volume of (25–50 µl) not greater than the size of the sample well.

Heat samples in boiling water for 5 min to ensure complete interaction between proteins and SDS. Load samples into the bottom of the wells. The samples can be conveniently loaded using a mircolitre syringe or a micropipette fitted with a narrow tip or a teflon tubing fixed to the end of a microlitre syringe.

Mark the position of the wells on the glass plate with a marker pen the presence of bromophenol blue in the sample buffer facilitate easy loading of samples. Load the well with standard marker protein in the sample buffer. Start electrophoresis at 100–125 V. After the dye has moved into the separating gel, increase to 200 V. The stacking gel helps in concentrating of the samples. The plates will become warm to touch, but should not be hot. If the plates are hot, lower the voltage. When the dye reaches the bottom of the gel, turn off the power pack.

Remove the gel plates and gently pry the plates apart. Use a spatuala or similar tool to separate the plates. Cut a corner from the bottom of the gel that is closer to the first well. Immerse the gel in staining solution overnight with uniform shaking. The protein binds the coomassie brilliant blue.

Transfer the gel to a suitable container atleast 200–300 ml destaining solution and place on a rocker. Dye that is not bound to proteins is thus removed. The destainer must be changed frequently, until the background of the gel is colourless. The protein bands are coloured blue. The destaining process can be stopped at a stage, when many bands are visualized as much as possible. Faint bands represent minute quantities of proteins. View through a light box and score the bands for further analysis.

Reference

Laemmli, U.K. (1970), Nature, 227: 680.

9.17 WESTERN BLOTTING

Western blotting involves the transfer of proteins from completed polyacrylamide gel into a nitrocellulose membrane.

Reagents

Transfer buffer (pH 8.3 – 8.4)

Tris base	18.2 g
Glycine	86.5 g
Methanol	1200 ml
Distilled water	4 litres

Make upto 6 litres with water.

Procedure

Separate the samples by electrophoresis in 12% polyacrylamide gels as described earlier. Equilibrate the gels after electrophoresis with transfer buffer for 5 min in trough. The Whatman filter papers should also be equilibrated in the transfer buffer for some time. Cut a piece of nitrocellulose paper to the size of the gel and immerse in the buffer. On one lid of the transfer cassette, first place three bits of filter paper and then place the gel carefully on the filter paper. Place the nitrocellulose paper carefully on the gel without entrapping air bubbles between the gel and the nitrocellulose paper. Place another set of filter papers on the nitrocellulose paper. Close the cassette with another lid and place in the transfer tank filled with transfer buffer. Electroblot the sandwich with the gel side facing the cathode and nitrocellulose paper side facing the anode. Carry out the transfer overnight at 20 mA in a cold room.

Reference

Burnette, W.N. (1981), Anal. Biochem. 112: 195.

Food Analysis

QUANTITATIVE ANALYSIS

Proximate analysis: this is the determination of the major constituents of food. These constituents include parameters such as (i) moisture, (ii) fat, (iii) ash, (iv) crude protein (v) carbohydrates and (vi) minerals.

10.1 DETERMINATION OF MOISTURE CONTENT

Food sample is dried at 125°C for 3 h in a hot air oven. The loss in weight is reported as percent moisture.

Procedure

Weigh accurately, 4-5 g of food material and transfer it into a preweighed moisture dish. This should be spread evenly. Dry the sample for 3 h at 125°C in a hot air oven. After drying, place the lid of the moisture dish and transfer the dish into a dessicator to cool the contents. Weigh the moisture dish rapidly, after 30 min of dessication and record the weight. Calculate the % moisture content by the following equation:

Calculation

$$\%\ \text{moisture} = \frac{(b\text{-}c) \times 100}{a}$$

where

a = Weight of the sample in grams

b = Weight of the dish + sample before drying

c = Weight of the dish + sample after drying

$(b\text{-}c)$ = Drop in weight due to loss of moisture

10.2 DETERMINATION OF OIL CONTENT IN FOOD

Oil content of the food is determined by drying the sample and extracting the oil using a Soxhlet apparatus. The solvent used to extract the oil is removed by evaporation and the residue is weighed and reported as percent oil.

Procedure

Accurately weigh 6 g of the food sample and grind it into a fine powder, using a mechanical blender. Transfer the sample into a moisture dish and dry the sample at 125°C for 2 – 3 h. Using a thimble, subject the sample to Soxhlet extraction. Set the heating mantle temperature in such a way that at least 160 –180 drops/min of solvents flow on to the pouch or the thimble. Continue the extraction for 6 – 8 h. After completion of the extraction, cool and disconnect the extraction flask. Transfer the contents of the flask quantitatively into a clean, dry and preweighed 250 ml glass beaker. Place the beaker on a hot water bath and evaporate the solvent in a fume hood. Dry the residue in a hot air oven at 105°C for 30 min to remove traces of solvent and moisture. Cool the beaker in a dessicator, weigh the contents and record the weight.

Calculation

$$\text{Oil content (\%)} = \frac{(a - b) \times 100}{c}$$

where,

a = Weight of the container + oil

b = Weight of the container

c = Weight of the sample in grams

10.3 DETERMINATION OF ASH CONTENT OF FOOD

The ash content of the food represents the inorganic constituents after the organic and volatile materials have been oxidized completely during the process of incineration at 600°C in a muffle furnace.

Procedure

Weigh 3 g of defatted food sample into a crucible of known weight. Set the temperature of the muffle furnace to 600°C. Place the crucible containing the food sample into the muffle furnace after attaining the set temperature, using a metal tong. Incinerate the sample for 2 h. Transfer the crucible into a desiccator and cool to room temperature. Immediately, weigh the crucible along with ash, to prevent moisture absorption. Record its weight. Repeat the process of incineration until a constant weight in obtained.

Calculation

$$\% \text{ Ash content} = \frac{(b - c) \times 100}{a}$$

where,

a = Weight of the sample in grams

b = Weight of the crucible + constant weight of ash

c = Weight of the empty crucible

$(b - c)$ = Weight of ash.

Reference

FAO Food and Nutrition Papers: 14/7: Manuals of Food Quality Control, FAO, UN, Rome, Italy, 1986.

10.4 ESTIMATION OF TOTAL PROTEINS BY MACRO-KJELDAHL'S METHOD

This is a method of choice for estimating protein content of food material, agricultural and clinical samples. This is also used to estimate the non-protein nitrogen of a sample, after precipitating out the proteins.

The estimation of nitrogen is done by Kjeldahl method which depends up on the fact that organic nitrogen when digested with sulphuric acid in the presence of a catalyst (selenium oxide, mercury or copper sulphate) is converted into ammonium sulphate. Ammonia liberated by making the solution alkaline is distilled in to a known volume of a standard acid, which is then back titrated. The protein content is obtained by multiplying the nitrogen value with 6.25.

Reagents

1. Digestion mixture: 98 parts K_2SO_4 and 2 parts $CuSO_4$.
2. 40% NaOH
3. N/10 NaOH
4. N/10 H_2SO_4
5. Methyl red indicator: 0.1g of the indicator dissolved in 60 ml of alcohol and water to make up to 100 ml.

Procedure

Weigh the sample (0.5 – 2.0g) into a dry Kjedahl's flask. Add 5 g of digestion mixture and 20 ml of pure conc. H_2SO_4 to the sample and digest the mixture by heating for 4 – 5 hours. Add glass beads to the reaction mixture to prevent bumping. After the contents of the flask become clear, continue the digestion for at least one hour. Cool the contents of the Kjeldahl flask, dilute with distilled water and add excess of 40% NaOH (about 75) to make the mixture alkaline.

Add a small quantity of pumice powder to prevent bumping during distillation. Ammonia is liberated during this process; collect the liberated ammonia liberated into a receiver containing 25 ml of N/10 H_2SO_4. Back titrate the excess of acid in the receiver against N/10 NaOH using 3 drops of methyl red indicator. Prepare a reagent blank similarly. Subtract the titre value of the blank from the value obtained for the sample to get the true titre value.

Calculation

If '*a*' g of the sample are taken and if '*b*' and '*c*' ml of alkali of normality '*d*' are required for back titration and to neutralize 25 ml of N/10 H_2SO_4 respectively, then the protein content in/100g of sample is

$$\frac{(c-b) \times 14d \times 16.25}{a \times 1000} \times 100$$

True nitrogen can be determined by subtracting NPN (which is estimated by precipitating the protein in the sample with TCA, tungstic acid or $Cu(OH)_3$ and determining the residual nitrogen in the protein free filtrate) from the total nitrogen.

10.5 NON-PROTEIN NITROGEN (NPN)

The non-protein nitrogen in foods include nitrogenous bases of nucleic acids, non-protein amino acids (eg. Ornithine, homocysteine, citrulline, β alanine, diamino propionic acid, desmosine etc., amino sugars and their derivatives, phosphor- and sphingolipids etc)

Proteins present in the food sample are precipitated by tricholoroacetic acid (TCA) and the non-precipitated nitrogenous components in the supernatant are estimated by salicylate-nitroprusside method.

Reagents

1. Tricholoroacetic acid 10% w/v.

Procedure

Weigh 1 g of defatted food sample and transfer it into a 20 ml graduated and stoppered centrifuge tube. Add 15 ml of TCA reagent and mix for 1 h. After 1 h remove the stopper and centrifuge the contents at 1500 rpm for 15 min. Carefully collect the supernatant into a 25 ml volumetric flask. Wash the residue twice with 5 ml TCA reagent and repeat the centrifugation step. Pool the TCA extract and make up the final volume to 25 ml with TCA reagent.

For the determination of non-protein nitrogen, take aliquots of 5 and 10 ml of TCA extracts. Determine the nitrogen content of the sample after digesting the extract with sulphuric acid and carry out as stated for protein. The values are expressed as percentage.

Reference

Singh, U. and Jambunathan, R.J. (1981), J. Agric. Food Chem. 29: 423.

Total carbohydrates, minerals and vitamins are estimated in the sections mentioned elsewhere in the book.

10.6 DETERMINATION OF RANCIDITY OF EDIBLE OILS – KRIESS TEST

Rancidity in oils is a complex process brought about by air and microorganisms. Rancid oil reacts with pholoroglucinol (1,3,5-trihydroxybenzene) under acidic condition to yield a pink colour, which is measured at 545 nm.

Reagents

1. Acid reagent: Trichloroacetic acid (30% w/v) in glacial acetic acid
2. Pholoroglucinol (1%w/v) in glacial acetic acid
3. Ethanol (95%)
4. Chloroform

Procedure

Take 5 ml of test oil sample and add 5 ml of chloroform in a stoppered boiling test tube and mix. Add 10 ml of acid reagent followed by 1 ml of pholorglucinol reagent. Incubate the tube at 45°C in a hot water bath for 15 – 18 min. after incubation add 4 ml of ethanol (95% v/v) and immediately record the absorbance at 545 nm.

Interpretation

1. An absorbance value of ≥ 1.0 indicates that the sample is highly rancid.
2. An absorbance value of > 0.2 indicates that the sample has incipient rancidity.
3. An absorbance value of ≤ 0.15 indicates no rancidity.

Rancidity can also be confirmed by determining the levels of free fatty acids and peroxide value.

Reference

Bedford, C.L. and Joslyn, M.A. (1937), Food Res. 2: 455.

Vitamins and Minerals

11.1 ESTIMATION OF ASCORBIC ACID IN LEMON JUICE (TITRIMETRIC METHOD)

Vitamin C (ascorbic acid) undergoes oxidation in air to for dehydro ascorbic acid. It is therefore able to reduce the dye 2,6-dichlorophenol indophenol, (in the oxidized form the dye is blue and undergoes decolouration on reduction).

Reagents

1. 2,6 dichlorophenol indophenol solution: Dissolve 52 mg of sodium salt of the dye and 42 mg of sodium bicarbonate in water. Make upto 500 ml in a volumetric flask.
2. Standard ascorbic acid: Dissolve 10 mg of ascorbic acid in 6% metaphosphoric acid and make up the final volume to 1 litre.

Procedure

Dilute the lemon juice 10 fold with 6% metaphosphoric acid. Pipette out 20 ml of the juice and titrate against the dye taken in the burette, until the dye turns light pink in colour. Note the amount of dye consumed. Similarly process the standard ascorbic acid solution.

Calculation

Concentration of ascorbic acid present in standard solution = 10 μg/ml

Amount of ascorbic acid in 20 ml of standard solution = 200 μg

If *y* ml of dye is required to oxidize 20 ml of standard solution, then 1 ml of the dye causes oxidation of 200/y μg of ascorbic acid.

If 20 ml of diluted lemon juice is oxidized by *z* ml of the dye then

Amount of ascorbic acid present in 20 ml of diluted juice $= \frac{200}{y} \times z$ µg

Ascorbic acid present in 100 ml of diluted juice

$$= \frac{200 \times z}{y} \times \frac{100}{20} \times \frac{1}{1000}$$

$$= \frac{z}{y}$$

∴ Ascorbic acid in 100 ml of undiluted juice $= \frac{z \times 10 \text{ mg}}{y}$

Reference

Harris, L.J. and Ray, S.N.(1935), Lancet, I, 71: 462.

11.2 DETERMINATION OF THIAMINE IN CEREALS AND FOODS

Thiamine undergoes oxidation with mild oxidizing agents such as ferricyanide to form thiochrome which is a fluorescent compound.

Reagents

1. **Sulphuric acid (0.1 N):** 0.1 ml of conc. Sulphuric acid is made upto 40 ml with distilled water.
2. **Potassium ferricyanide (0.03%):** 30 mg of potassium ferricyanide and dissolve in 100 ml of distilled water.
3. Iso – butanol
4. Sodium sulphate (anhydrous)
5. **Sodium hydroxide (15%):** Dissolve 15 g of NaOH in water and make the final volume to 100 ml.
6. **Stock thiamine solution:** 5 mg of thiamine hydrochloride in dissolved in 100 ml of water.

   Working standard: Take 2 ml of the stock and dilute it to 100 ml with water. This solution contains 1 µg thiamine hydrochloride per ml.

Procedure

Grind 1 – 2 g of food material with 20 ml of 0.1 N H_2SO_4 using a mortar and pestle. Centrifuge the suspension at 4000 rpm for 20 min. Decant the supernatant and repeat extraction with 10 ml of acid. Collect the supernatant in a volumetric flask and make upto to 50 ml with 0.1 N H_2SO_4. Pipette a small volume of the thiamine extract and add 3 ml if potassium ferricyanide reagent followed by 3 ml of 15% NaOH. Shake the contents for 15 min. add 10 ml of iso-butanol and shake the mixture thoroughly. Allow it to stand till the two phases separate out completely. Discard the lower aqueous phase and remove the traces of water in the upper organic phase by adding a small amount of solid anhydrous sodium sulphate. Pipette out graded volumes of thiamine standard (0 – 5 μg) and process similarly. The fluorescence intensity of the sample and standards are recorded using a primary filter of 365 nm and secondary filter of 435 nm. The amount of thiamine is expressed as mg/100 g of sample.

11.3 ESTIMATION OF RIBOFLAVIN IN URINE SAMPLE

Riboflavin is extracted in the butanol-pyridine mixture and fluorescence intensity is measured using a fluorimeter.

Reagents

1. Potassium permanganate – 5% w/v
2. Hydrogen peroxide – 3%
3. Butanol – pyridine mixture (92:8)
4. Sodium sulphate (anhydrous)
5. Glacial acetic acid
6. Riboflavin standard: Dissolve 25 mg of riboflavin in 500 ml of distilled water, add 1.2 ml of glacial acetic acid and make the volume to 1 litre. Dilute 5 ml of this solution to 50 ml with distilled water to get a working standard containing 2.5 μg riboflavin/ml.

Procedure

Take 2 ml of urine sample in a test tube and add 0.5 ml of glacial acetic acid. Then add 0.5 ml of potassium permanganate and shake well and leave it for a minute. Add 0.5 ml of 3% H_2O_2 until a purple colour of excess potassium permanganate disappears completely. Then add 10 ml of butanol-pyridine mixture and shake well for one minute. Leave the tubes undisturbed for a few minutes until the two layers separate out. Discard the aqueous layer. Add 1g of sodium sulphate to remove excess moisture. Shake and keep in the dark for 10 – 15 min. Read the fluorescence intensity at 450 nm excitation light and 545 nm emission light. Process the standards similarly and read its fluorescence.

Calculation

The fluorescence intensity of sample = *a*; and

The standard containing y μg riboflavin = *b*

The volume of urine taken for the test = *v* ml

Then, the riboflavin content (mg/ml) in urine is $= \frac{a}{b} \times \frac{y}{v}$

11.4 MICROBIAL ASSAY FOR ESTIMATION OF NIACIN IN FOOD

Niacin can be determined microbiologically using *Lactobacillus plantarium.* The organism provides highly specific analysis and responds equally to nicotinc acid and nicotinamide but not to intermediate precursors of nicotinic acid such as picolinic and quinolinic acids. Even though *Lactobacillus* requires an exogenous supply of pantothenic acid, this complication is overcome by using a culture medium which contains sufficiently high amounts of pantothenic acid so that growth of the organism is dependent exclusively on niacin content of the test material. Niacin from the food material is extracted by suspending the sample in H_2SO_4 and then autoclaving. After filtration and precipitation of protein at pH 6.8, the filtrate is diluted to get niacin concentration of about 0.1 μg niacin/ml. the assay organism is then grown in basal medium under specified conditions in presence of standard concentrations of niacin and sample extract. Growth of the microorganism is measured by monitoring the turbidity at 660 nm. By comparing growth rates in media containing the sample extract with those in the presence of graded amounts of this vitamin the niacin content in the sample can be determined.

Reagents

1. **Niacin stock solution:** Dissolve 50 mg of niacin in 500 ml of 50% alcohol to give a solution containing 100 μg niacin/ml.
2. **Working standard:** Dilute the stock solution to obtain 0.1 μg of niacin/ml.
3. **Acid hydrolysed casein:** To 100 mg of casein, add enough 95% ethanol and reflux the mixture for 2 – 3 h and filter. Collect the residue and repeat this step twice. Remove the alcohol from the refluxed residue at room temperature under vacuum; add 250 ml of conc. HCl and 250 ml of distilled water and reflux for 8 – 10 h. Distill off HCl under vacuum below 80 °C. Remove HCl completely by repeating the process 2 – 3 times. Dissolve the hydrolysed casein in 600 ml distilled water and adjust the pH to 3.5 with 40% NaOH and filter the preparation.
4. **L-Cystine:** Add 4.0 g in a small volume of water and dissolve it using minimum volume of HCl and make up the volume to 500 ml using distilled water.
5. **DL-tryptophan:** Dissolve 2 g in water using a few drops of liquor ammonia. Adjust the pH to 3.5 and make the volume to 500 ml using distilled water.
6. **Biotin:** Dissolve 25 mg in 500 ml of 50% ethanol.
7. **Calcium pantothenate:** Dissolve 54.4 mg of calcium pantothenate in distilled water.

8. **Vitamin solution:** Dissolve 20 mg riboflavin, 10 mf thiamine hydrochloride, 10 mg p-amino benzoic acid, 40 mg pyridoxine HCl in 1 litre of 0.02 N HCl.
9. **Adenine, Guanine, Uracil (AGU) solution:** Prepare a solution containing 0.2 g each of adenine sulphate, guanine hydrochloride and uracil in distilled water using a few drops of conc. HCl to aid dissolution and make up to 100 ml with distilled water.
10. **Salt solution A:** Make a salt solution containing 25 g each of K_2HPO_4 and KH_2PO_4 in 250 ml water.
11. **Salt solution B:** Dissolve the following salts in distilled water and make the volume to 250 ml. Add few drops of HCl to obtain a clear solution.

$MgSO_4 . 7H_2O$	10.0 g
NaCl	0.5 g
$FeSO_4 . 7H_2O$	0.5 g
$MnSO_4 . 4 H_2O$	0.5 g

12. **Basal medium:** Prepare the basal medium by mixing the specified volume of the various solutions mentioned below:

Casein hydrolysate	10 ml
DL-tryptophan	5 ml
L-Cystine	5 ml
AGU solution	1 ml
Vitamin solution	2 ml
Salt solutions (A & B)	1 ml each
Calcium pantothenate	0.2 ml
Biotin	0.8 ml

Then add 2 g of glucose (anhydrous) and 3.32 g of $CH_3COONa.3H_2O$ to the above mixture and adjust the pH to 6.8 and make the final volume to 100 ml. filter the mixture through Whatman No.1.

Procedure

Weigh sufficient amount of powdered food so as to contain approximately 0.1 mg niacin and transfer it to a 100 ml conical flask. Add 50 ml of 1N H_2SO_4. Autoclave the mixture for 30 min at 15 psi and adjust the pH to 4.5. After cooling, filter it through Whatman No.1 and make up the volume to 100 ml. Take suitable volume of the aliquot and dilute it so that it contains 0.1 μg/ml niacin.

To 15 ml test tubes, take varying volumes from 0.5 to 3.0 ml of the sample extract. For each volume of the aliquot, take atleast two test tubes. Make the final volume of the contents in the test tubes to 5 ml with water. Then add 5 ml of the basal media to each tube. Plug the tubes with nonabsorbent cotton.

Similarly prepare in duplicate, test tubes containing 0, 0.5, 1.0, 1.5, 2.0, 2.5, 3.0, 3.5, and 4.0 ml of standard solution which contains 0.1 μg of niacin/ml. add appropriate volume of water to each of the tubes to make the final volume to 5.0 ml followed by addition of 5 ml if the basal media and closet he tubes with a cotton plug.

Autoclave the tubes containing the sample aliquot as well as the standards at 15 psi for 15 min. Allow the tubes to cool and pipette aseptically 0.1 ml of the inoculum of the broth. Incubate the tubes at 37°C for 18 h and record absorbance of each tube at 660 nm.

Calculations

Prepare the standard curve by plotting a graph of A_{660} *vs* ng niacin for tubes containing the reference standards. From the standard curve, determine the niacin content in each of the sample tubes. Calculate the amount of niacin per 100 g of food sample by taking into consideration the dilution factors involved during preparation of the sample extract and the amount of aliquot taken for estimation of the vitamin.

MINERAL

Sample processing for mineral estimation

- Nitrogen
- Phosphorus
- Calcium and magnesium
- Magnesium
- Chloride
- Copper
- Iron – 2,2'-dipyridyl method
- Bathophenanthroline method

11.5 SAMPLE PROCESSING FOR MINERAL ESTIMATION

There are three methods which have generally been employed for processing the sample prior to mineral analysis.

1. Direct solutions

Liquids samples can be analyzed directly without any pretreatment after dilution with water or some other suitable reagent. Serum of instance is diluted 5 times with 80% (v/v) butanol indeionixed water for mineral determination. Fruit juices and beverages such as tea, fresh fat-free milk and wine can similarly be analyzed after their proper dilution.

2. Ashing

The entire organic matter is destroyed and the non-combustible material is recovered as ash. The minerals are then extracted from the ash with an acid (usually dil HCl), filtered and diluted to a known volume with deionized water and finally estimated quantitatively. In this method the organic compounds in a sample is decomposed by incineration at high temperatures (500 – 600°C) for 4 – 12 h using a muffle furnace.

Procedure

Take a known weight of the sample in a previously weighed silica crucible. Heat it first on a Bunsen burner on a low flame till the contents are charred. Now transfer the crucible to a muffle furnace and heat at 500 - 600°C till it is completely converted to white ash. Remove the crucible and keep it in a dessicator till it cools. Weigh the crucible and calculate the weight of the ash. Moisten the ash with a small amount of glass distilled water (0.5 – 1.0 ml) and add 5 ml of dilute HCl (1 : 4) to it. Evaporate the mixture to dryness on a boiling water bath and then add another 5 ml of HCl and evaporate the solution t dryness again. Now add 4 ml of HCl and few ml of water and warm the solution on a boiling water bath. Filter the extract into a 100 ml volumetric flask using Whatman No. 40 filter paper, allow to cool and make up the volume to 100 ml in a standard flask. Aliquots of this extract is taken for the estimation of minerals.

3. Wet Digestion

Organic matter in plant or animal samples can be oxidized and destroyed by prolonged boiling with mixtures of H_2SO_4, $HClO_4$ and HNO_3. A quite frequently used mixture is that containing HCl and 70% $HClO_4$. The main advantage of the wet digestion procedure is that it almost totally eliminates elemental losses by volatilization as it is carried out at relatively low temperatures. However, it requires continued attention of the analyst, is tedious and also susceptible to reagent contamination.

Procedure

Weigh 3 g dried and finely ground sample and transfer it to a 100 ml Kjeldahl flask. Add 25 ml of a mixture of conc. HNO_3, conc $HClO_4$ and conc. H_2SO_4 (3:2:1) and shake well ensuring that no dry lumps are left behind. Drop a clean acid washed glass bead into the flask to prevent bumping during digestion.

Allow it to stand overnight in a fume cupboard. Heat it on a digestion heater and watch out for foaming during the first hour especially in samples with high fat content. Allow it to cool, filter the digested sample through Whatman filter paper into a 100 ml volumetric flask. Make sure that the contents are transferred in the standard flask by washing with deionized water. Wash the residue of silica on the filter paper with very dilute HCl (1:19) diluted with water in order to wash down the slats completely. Make up the volume of the collected filtrate to 100 ml.

Nitrogen

In solution ammonium ions gives a reddish yellow colour with Nesseler's reagent in alkaline medium. The colour intensity is proportional to the concentration of ammonium ions and is measured.

Reagents

1. Nessler's reagent:

Solution A: 7.5 g of potassium iodide is dissolved in 100 ml of water. 10g of mercuric iodide is added.

Solution B: 10 % sodium hydroxide 50 ml of solution A and 35 ml of solution B are mixed in a beaker and made up to 100 ml.

Stock standard: 94.28 mg of ammonium sulphate is weighed and made upto 100 ml with water. 10 ml of the stock standard is diluted to 100 ml with water.

Procedure

Pipette out 1 ml of the sample and standards in the range of 10 – 50 µg. Make up the volume in all the tubes to up to 9.5 ml with water. Add 0.5 ml of Nessler's reagent. The yellow colour developed is read at 540 nm.

11.6 PHOSPHORUS

Inorganic phosphorus forms phosphomolybdate with molybdate solution. This compound is reduced by aminonaphthol sulphonic acid to give a blue colour, the intensity of which is proportional to the concentration of phosphate present. After 20 min the intensity of colour is read at 620 nm.

Reagents

1. ***Amino-naphthol sulphonic acid:*** Weigh 0.5g of aminonaphthol sulphonic acid. To this add 195 ml of 15% sodium bisuphite and 5 ml of 20% sodium sulphite solution was added and warmed.
2. ***Ammonium molydate solution:*** Dissolve 2.5 g of ammoium molybdate in 100 ml of distilled water.

3. ***Standard phosphorus solution:*** Dissolve 35.1 mg of potassium dihydrogen phosphate in 100 ml of distilled water. Working standard is prepared by diluting 10 ml of stock to 100 ml with distilled water.

Procedure

Pipette out 1 ml of test sample and make it upto to 9 ml with deionized water. Then add one ml of ammonium molybdate and 0.4 ml of ANSA. Incubate the tubes at 37°C for 10 min and the colour developed is read at 640 nm. Phosphorus standard in the range 4 – 20 µg is processed in the same manner.

Reference

Fiske, C.H. and Subbarow, Y.J. (1925), Biol Chem. 66: 375.

11.7 CALCIUM

The ionic calcium present in aqueous medium is precipitated as calcium oxalate by treating with ammonium oxalate. The amount of calcium is determined titrimetrically by using potassium permanganate solution in an acidic medium.

Reagents

1. Ammonium oxalate – 4%
2. Ammonium hydroxide – 2%
3. Potassium permanganate (0.01N)
4. Sulphuric acid – 1N
5. Concentrated sulphuric acid
6. Oxalic acid (0.01 N)

Procedure

Pipette out 25 ml of oxalic acid into a 250 ml conical flask. Add 1 ml of conc. sulphuric acid and heat the contents to 70°C on a waterbath. Titrate immediately against potassium permanganate taken in a burette, while the contents in the conical flask are still hot. During the initial titration, add potassium permanganate solution drop wise with constant stirring until the end-point a pale permanent pink colour is obtained. Repeat the titration for concordant values and from this calculate the normality of potassium permanganate.

Take 2 ml of the sample solution into a graduated glass centrifuge tube and add 2 ml of deionised water followed by 2 ml of ammonium oxalate and vortex gently. Leave the tube for 30 min at room temperature. Centrifuge the contents at 2500 rpm for 10 min. Discard the supernatant without disturbing the precipitate. Drain the residual supernatant by inverting the tube onto a blotting paper. Wash ther precipitate, thrice with 3 ml of 2% ammonium hydroxide solution and vortex gently. After each wash, centrifuge the content as given above. The washing step is essential in order to ensure complete removal of ammonium oxalate.

To the washed precipitate, add 2 ml of 1 N H_2SO_4 and mix. Place the tubes in a boiling water bath, until all the precipitate is dissolved and then titrate the hot solution with standardized potassium permanganate until the end point a pale permanent pink colour is reached. Conduct a blank titration using 2 ml of water and 2 ml of 1 N H_2SO_4 only.

Calculation

Sample titre value (ml) – Blank titre value (ml) = X

One ml of 0.01 N potassium permanganate = 0.2004 mg of calcium

$$\text{Calcium (mg \%)} = \frac{X \times 0.2004}{2} \times 100$$

Reference

Clark, E.P. and Collip, J.B. (1925), J. Biol. Chem. 63: 461.

11.8 MAGNESIUM

Titan yellow gives a red colour with magnesium.

Reagents

1. Sodium tungstate – 10%
2. Sulphuric acid – 2/3N
3. Gum ghatti – 0.1%. Suspend 0.1g of powdered gum ghatti in a muslin bag in 100 ml of distilled water for 24 hours.
4. Titan yellow- 0.05%
5. Sodium hydroxide – 4N
6. **Stock standard solution:** Dissolve 8.458 g of $MgCl_2.6H_2O$ in distilled water and make up to 1 litre.
7. **Working standard:** Dilute 1 ml of stock solution to 200 ml with water.
8. Calcium chloride solution 0.05mg calcium per ml. Dissolve 13.88 mg of calcium chloride in water and make up to 100 ml.

Procedure

Dilute 1 ml of the sample with 5 ml of distilled water and precipitate the proteins by adding 2 ml of 10% sodium tungstate and 2 ml of 2/3N sulphuric acid and centrifuge. To 5 ml of the supernatant add in turn 1 ml of distilled water, 1ml of the gum ghatti, 1ml of 0.05% titan yellow and 2 ml of 4N NaOH. At the same time put up 1 ml of calcium chloride and 5 ml of water, 1 ml of calcium chloride and 2.5 ml of the standard for use and 2.5 ml of water for blank, process these tubes in the same way as the test. Read the standard and unknown against the blank using a green filter at 520 nm.

Reference

Neill, D.W., and Neely, R.A. (1956). J. Clin. Path., 9:162.

11.9 CHLORIDE

The chloride present in the aqueous medium is precipitated by excess of silver nitrate to silver chloride in the presence of nitric acid. The unreacted silver nitrate is determined titrimetrically with ammonium thiocyanate, using ferric ammonium sulphate as an indicator to red coloured end point.

Reagents

1. Silver nitrate solution: Dissolve 16.988 g of silver nitrate in 400 – 500 ml of double glass distilled water. Separately dissolve 40 g of ferric ammonium sulphate in a 500 ml glass beaker, containing 200 ml of double distilled water and 85 ml of conc. nitric acid. Mix this solution and make upto 1 litre with double distilled water.
2. Ammonium thiocyanate – 0.9% w/v

Procedure

Take 10 ml of silver nitrate solution in a 100 ml conical flask and titrate this against ammonium thiocyanate solution taken in a burette, to a pale red colour. 10 ml of silver nitrate solution should consume 10 ml of ammonium thiocyanate solution (0.9%). In case the silver nitrate solution consumes less than 10 ml of ammonium thiocyanate, dilute the latter accordingly in such a manner that 1 ml of silver nitrate consumes 1 ml of ammonium thiocyanate solution.

Dilute 10 ml of the sample solution with 25 ml of double-glass distilled water in a 100 ml volumetric flask. Add 20 ml of silver nitrate solution and mix. Make up the volume to 100 ml with double-glass distilled water. Filter this solution using a glass funnel plugged tightly with glass wool or Whatman No 1 filter paper. Take 50 ml of the filtrate into a 250 ml conical flask and titrate against ammonium thiocyanate solution taken in a micro-burette until a pale red colour is obtained as the end point.

Calculation

Chloride (mEq/L)

$$= \frac{10 \text{ ml } (AgNO_3 \text{ solution}) - X\text{ml (titer value)} \times 0.1 \text{ (mEq)}}{5 \text{ (sample volume)}} \times 1000$$

References

1. *Van Slyke, D.D. (1923), J. Biol Chem. 58: 523.*
2. *Day, R.A. Jr and Underwood, A.L. (1993), Quantitative Analysis 6th ed. Prentice-Hall of India (P), Ltd. New Delhi.*

11.10 COPPER

The copper is released from its linkage to proteins by means of hydrochloric acid, the proteins precipitated by trichloroacetic acid, and the copper extracted from the protein-free fluid forms a golden yellow complex with sodium diethyldithiocarbamate.

Reagents

1. Hydrochloric acid – 0.1N
2. Hydrochloric acid – 6N
3. Sodium pyrophosphate, 6%
4. Ammonia
5. Sodium diethyldithiocarbamate – 0.4%
6. Amyl alcohol – ether mixture 1:1
7. Sodium sulphate
8. **Stock standard:** Dissolve 0.398 g of copper sulphate in water, add 0.1 ml of concentrated sulphuric acid and make up to 1 litre. This solution contains 100 µg per ml.
9. **Working standard:** Dilute 1 ml of stock to 100 ml.

Procedure

To 3 ml of the sample add 1 ml of 0.1 N HCl and warm in a boiling water bath, stirring continuously until the mixture begins to cloud. Cool. Then add 1.5 ml of 6N HCl and stand for 10 min. add 3 ml of 20% TCA, mix well and centrifuge. Decant the supernatant fluid and wash the precipitate with 3 ml of 5% TCA. Centrifuge again and combine the supernatant fluids. Add 1 ml of 6% sodium pyrophosphate and 2 ml of ammonia. Add 1 ml of sodium diethyl dithiocarbamate and shake well for about 5 minutes with 5 ml of the amyl alcohol – ether mixture to extract the copper. Remove the amyl alcohol layer and dry by shaking with a little powdered anhydrous sodium sulphate. Read using a violet filter at 440 nm. Process a standard containing 1 – 5 µg /ml and treat the same way as the test.

Reference

Eden, A. and Green, H.H. (1940), Biochem. J. 34: 1202.

11.11 ESTIMATION OF IRON

Ferrous iron gives a pink colour with 2,2'-dipyridyl. A solution of dipyridyl in acetic acid is added to the sample followed by a reducing agent.

Reagents

1. 0.1% 2,2' dipyridyl in 3% acetic acid (v/v)
2. **0.1 M sodium sulphite:** 1.26 g of anhydrous sodium sulphite or 2.52 g of $Na_2SO_3 \cdot 7\ H_2O$ dissolved in water and made up to 100 ml. Prepared fresh.
3. Chloroform
4. Standard solution containing 100 μg iron /ml. Dissolve 0.498 g of $FeSO_4 \cdot 7\ H_2O$ in water, added 1.0 ml conc H_2SO_4 and made up to a liter.
5. **Working standard:** Dilute 3.0 ml of the stock solution to 100 ml with water obtain a solution containing 3 μg/ml.

Procedure

Take 0.4 ml each of sample, 0.1 M sodium sulphite and of dipyridyl reagent in a glass stoppered tube, mix well and centrifuge. Heat the mixture in a boiling water bath for 5 min. Cool the tubes to room temperature and add 1.0 ml chloroform, stopper the tubes and shake well for 30 seconds. Remove the stopper centrifuge for 5 min at 3000 rpm.

The colour extracts into the organic phase, aspirate this phase and measure the absorbance at 520 nm against water blank. Working standards containing 3 μg/ml was used. Values are expressed as mg/g.

Reference

Ramsay, W.N.M. (1954), Biochem. J. 57:xvii.

11.12 ESTIMATION OF IRON USING BATHOPHENANTHROLINE

Bathophenanthroline, 4,7-diphenyl-1,10-phenanthroline, gives a more intense colour than dipyridyl.

Reagents

1. Thioglycollic acid - 80%.
2. Hydrochloric acid
3. Sodium acetate - 58%
4. Tricholoroacetic acid - 30%
5. Bathophenanthroline - 0.02% in isopropanol

6. Standard solution: Dissolve 10 mg iron in100 ml of water and dilute 2 ml of stock to 100 ml to get a working standard.

Procedure

Pipette 2 ml of sample in to a test tube, 3 ml of water, 1 drop of conc HCl and 1 drop of thioglycollic acid. Mix and stand for 30 minutes. Set up a blank and standard at the same time with 2 ml of water and 2 ml of working standard respectively. To each then add 1 ml of 30% TCA, mix, and allow to stand for about 10 minutes and centrifuge. To 3 ml of the supernatant add 0.4 ml of sodium acetate and 2 ml of bathophenenthroline. Mix and read after 5 minutes using a green filter at 540 nm. Results are expressed as mg/l.

Reference

Peters, J.P. and Van Slyke, D.D. (1931), Quantitative Clinical Chemistry, Vol II, Methods, Bailliere, Tindall and Cox, London.

Chapter 12

Haematology

12.1 COLLECTION OF BLOOD SPECIMENS FROM PATIENTS

Capillary or venous blood is used for almost all estimations made on blood. Capillary blood is usually obtained from a finger or thumb and the most convenient place is on the thumb about half a centimeter from the side of the nail; the tip of the finger can also be used. Maximum number of hours allowed between the taking of blood and actual examination: haemoglobin and RBC is 24 hrs, but while performing fragility, sedimentation, corpuscles and packed cell volume it is 3 hours. Platelet count and blood films should not be done with stored blood.

Anticoagulants

1. Ammonium and potassium oxalate mixture

   General purpose anticoagulants of haematological investigations:

Ammonium oxalate	-	1.2 g
Potassium oxalate	-	0.8 g
Distilled water	-	100 ml

   Dissolve the oxalate salts in the distilled water and make up to 100 ml. deliver 0.4 ml of the solution with the help of a graduated pipette into small bottles or tubes marked at the 4 ml of level. Place the bottles or tubes in the oven at 80°C to evaporate the sater content from the oxalate solution. Close the bottle with screw caps or cork to tubes and labe as Wintrobe's bottle.

   Oxalated blood may be used for total RBC and WBC counts, haematocrit or PCV and haemoglobin estimation. Plasma may be used for estimation of bilirubin and prothrombin time. Smears are suitable for examination, only if prepared immediately, as the corpuscles undergo morphological changes. This anticoagulant should not be used for blood intended for chemistry.

2. **E.D.T.A (Ethylene diamine tetra acetic acid – disodium salt)**

EDTA	-	10 g
Distilled water	-	100 ml

Dissolve the disodium salt in the distilled water and make upto 100 ml mark. Deliver 0.08 ml of the solution with (1/10 marking) graduated pipette into small tube or bottle to 4 ml mark. Allow the water content of the solution to evaporate at room temperature.

12.2 HAEMOGLOBIN ESTIMATION

Sahl's acid haematin method

A simple but rather inaccurate method which estimates oxyhaemoglobin and reduced haemoglobin.

Place N/10 hydrochloric acid upto the mark 20 in the graduated tube. Add 20 cu mm of blood (0.02 ml) with the help of haemoglobinometer pipette. Rinse the pipette two to three times, mix well and allow to stand for five minutes till the solution becomes dark brownish colour. The solution is diluted drop by drop, each time mixing the solution with a stirring rod until it matches the standard. Read the result from the scale on the graduated tube by observing graduation mark at the lower edge of the meniscus at the top of the liquid column. If the final reading is not made within ten minutes, after mixing the blood with hydrochloric acid N/10, 2% should be deducted from the results obtained. In every anaemic case, the quantity of blood as also hydrochloric acid N/10 should be doubled and the result divided by two. It is best to match daylight.

Normal values

Men: 13.5 – 18 per 100 ml.

Women: 11.5 – 16.5 per 100 ml

Cyanmethaemoglobin Method

A reliable method for the estimation of total haemoglobin

Reagents

1. **Drabkin's solution:** dissolve 1.0 g sodium bicarbonate, 0.2 g potassium cyanide and 0.2 g potassium ferricyanide in 1 litre of distilled water.

Procedure

Expel 0.02 ml of blood into 5 ml of Drabkin's solution and rinse out the pipette several times with the mixture. Stopper the tube with rubber bung and mix by inverting several times. Allow it to stand for 10 min at room temperature. Match against standard with yellow-green filter in colorimeter.

The standard is prepared from blood of known haemoglobin content diluted 1:250 with Drabkin's solution. Standard solution with a haemoglobin content of 5 g, 140 g and 15 g in 100 ml are recommended.

12.3 TOTAL COUNT OF BLOOD CELLS

Red Blood Cells (Erythrocytes)

The mature erythrocytes is a biconcave disk, circular in shape, centrally unstained and periphery stained, pink in colour.

Reagent

RBC diluting fluid (Hayem's fluid): Dissolve 0.5 g sodium chloride, 2.5 g sodium sulphate, 0.25 g mercuric chloride in 100 ml of distilled water.

Procedure

Draw the blood from the finger directly into RBC pipette or from a sample of oxalated blood exactly to the 0.5 mark (Thoma pipette mark 101) and immediately draw up the diluting fluid mark 101 and rotate the pipette between the thumb and forefinger. This will give a dilution of 1:200.

Clean the counting chamber and cover glass thoroughly. Place the cover glass in position over the ruled area, using gentle pressure. Mix the suspension thoroughly by rotating the pipette for about a minute, holding it in horizontal position and finally shake sidewise. Expel the fluid from the stem of the pipette and without loss of time, fill the chamber by holding the pipette at an angle of 45 degrees and lightly touching the tip against the edge of the cover glass. Care should be taken to ensure that the suspension does not flow into the moats on either side, nor should any bubble form under cover glass. Allow two to three minutes for the red corpuscles to settle.

Count the number of RBC's in 80 small squares (4 squares of 16 at the four corners and one of 16 at center). Do not count the cells touching the lower and right hand lines, but count the cells touching the upper and left hand lines.

12.4 BULK DILUTION METHOD

In a clean test tube take 3.98 ml of RBC diluting fluid. Add 0.2 ml of blood with help of haemoglobin Sahli's pipette. Mix well. Now the dilution is 1 in 200, fill the counting chamber as usual, count and calculate.

Calculation: Count the number if cells in 80 small squares and add multiply by 10,000.

The area of small square is 1/400 sq.mm.

The depth of the counting chamber is 1/10 mm.

Therefore the volume of a small square is $1/400 \times 1/10 = 1/4000$ cu mm.

The dilution of the blood is 1/200.

$$\text{Total RBC} = \frac{\text{Cells counted}}{1} \frac{4000}{80} \frac{200}{1}$$

12.5 WHITE BLOOD CELL (LEUCOCYTES) COUNT

Reagents

WBC diluting fluid – Turcks fluid: Dilute 3 ml of glacial acetic acid with 97 ml of distilled water. Add gention violet to give a pale violet colour.

Procedure

Draw the blood upto the 0.5 mark in WBC pipette marked 11 and dilute upto the mark 11 with WBC fluid as described in RBC counting and fill the counting chamber in the same manner. Allow 3 minutes for cells to settle. If the Neubauer chamber is used count the cells in the four corner blocks. Each of these 4 square millimeter area is subdivided into 16 squares, by using the low power objective and a medium ocular. In counting the cells include those cells touching on the inner lines on the right and top but do not count the cells touching the lines on the left and bottom. The difference between the two square millimeter areas should not be more than 10 WBC's.

Bulk Dilution Method

In a clean test tube take 0.38 ml of WBC diluting fluid. Add 0.02 ml of blood with help of haemoglobin Sahli pipette. Mix well. Now the dilution is 1 in 20, fill the counting chamber as usual, count and calculate.

Calculation: Total number of cells counted in 4 squares/2 and multiply by 100 or,

To get WBC's per cu.mm

No: of WBC's counted in four corners	X cells
The volume of square is	1/10 cu. Mm
The blood was diluted to	1/20

Therefore the number of cells per cu.mm of undiluted blood

$$= \frac{X \times 10 \times 20}{4}$$

Differential Leucocyte Count

In a good blood film the corpuscles should be spread evenly, no Rouleax formation is to be seen. After general examination, of the film see if the following cells are present in the normal proportion of RBC, WBC, platelets and apart form this not if any immature cells or abnormal cells among the RBC or WBC and the size and shape of the red cells or any inclusions before proceeding to differential count.

Preparation of Film for Staining

Take a few perfectly clean grease-free slides. Clean the finger or ear lobe and prick and clean out the first drop of blood. Collect a big drop of blood on a clean glass slide and spread with a second slide. As soon as the blood has spread entirely across the end of the spreader slide (the spreader must be held at angle of 45 degree), with a quick movement push the spreader towards the other end of the under slide. The blood film should not be too thin or too thick. Allow the blood to dry.

Leishman's Method

A Roamnowsky type stain designed to differentiate leucocytes:

Leishman's stain powder	-	0.15g
Methyl alcohol, acetone free	-	100 ml

Take 0.15 g of Leishman's stain powder in a glass mortar add a few milli liters of methyl alcohol, grind dissolve the stain completely. Transfer the stain to a 100 ml measuring cylinder. Add little more methyl alcohol and grind, repeat the procedure the stain powder has dissolved and make up the volume to 100 ml, this stain is ready for use after two to three weeks.

The blood film is placed in a level position. The dry blood film is covered with the stain, which should be evenly distributed over the entire slide. At the end of one minute double the quantity of distilled water is carefully added and mixed with the stain by means of a clean pipette. The film is allowed to stain for seven or eight minutes and the excess stains is removed by washing with the distilled water for two minutes. The water is then washed off with distilled water. Dry the film in air, when the film is dried examine microscopically.

Results

Nuclei of Neutrophils – purple

Eosinophil granules – orange

Basophilic granules – dark blue

Lymphocytes – dark blue nuclei with pale blue cytoplasm

Platelets – violet granules

Auer bodies – red and

RBC – salmon pink.

12.6 PLATELET COUNT

Platelet in blood were counted using the improved Neubaur counting chamber by the method of Dacie and Lewis (1977).

Reagents

Dacie's fluid: This was prepared by dissolving 5 g of sodium citrate and 1 ml of 40% formaldehyde and making upto 100 ml with water. To 19 ml of this solution 1 ml of 0.2% brilliant crystal blue solution was added just before use. This solution was filtered and used.

Procedure

Venous blood collected with EDTA was used for platelet count 0.05 ml blood was diluted with 0.95 ml of Dacie's fluid and mixed well. Using a narrow bore Pasteur pipette, the counting chamber was filled with the diluted blood. The cells were allowed to settle to the bottom of the chamber for 15 min. To prevent from drying the chamber was placed in a petridish which contained a piece of wet filter paper.

Using 40 X objective with reduced condenser aperture, the platelets were counted in 1/5 sq mm. 5 ml of the small squares of the large centre square. From this the number of platelets in 1 cu mm of blood was calculated as

$$\frac{\text{Cells} \times \text{blood dilution} \times \text{chamber depth}}{\text{Area of chamber counted}}$$

Platelet count is expressed as number of cells $10^5/mm^3$.

12.7 RETICULOCYTE COUNT

Reagents

Reticulocyte diluting fluid

Brilliant cresyl blue	–	0.5g
Sodium citrate	–	0.4g
Sodium chloride	–	100 ml

Procedure

Place a small drop of reticulocyte diluting fluid at the end of a slide and allow it ot dry. Many slides can be kept ready for use. Prick the finger and place a drop of blood on the counting fluid. Mix the blood with the counting fluid using a glass rod and allow to stand for about 30 sec. Make a thin smear using the same technique for differential oil immersion objective. Count 1,000 erythrocytes, noting the number containing the bluish strands of reticulum. Divide this number by 10 to obtain the percentage of reticulocytes.

Normal - 0.5 to 1%

12.8 ABSOLUTE EOSINOPHIL COUNT

Reagents

1. Eosinophil diluting fluid
2. Mix 5 ml of 1% eosin (aqueous solution) and 5 ml of acetone and make up to 100 ml with distilled water. Filter and store.

Procedure

Make a dilution of 1 in 20 by adding 0.1 ml of blood to 1.9 ml of eosinophil diluting fluid in a clean bottle. Mix well. WBC pipette can be used for diluting the blood. Allow to stand for 15 minutes. Mix the solution thoroughly and charge the Fuchs Rosenthal counting chamber and stand for 5 minutes and allow the cells to settle down. Count all the cells in the entire ruled area (i.e area of 16 sq mm, volume 3.2 cu.mm).

Calculation

If N eosinophils are counted in 3.2 cu mm then the total number of eosinophils per cu.mm = $N \times 20$ (dilution) or $\frac{N \times 6.25}{3.2}$

12.9 BLEEDING TIME

The bleeding time is the time required for the blood to stop flowing from a deep cut in the finger. Rub the finger briskly to ensure adequate circulation, clean and puncture deeply. Note the time when the first drop of blood appears. Blot the blood with a filter paper every 30 sec without touching the skin. Note the time when bleeding stops the interval between the appearance of the first drop and the removal of the last drop represents the bleeding time.

Normal: Bleeding time by this method is 1 to 3 min.

12.10 COAGULATION TIME

Capillary tube method

Prepare 3 capillary tubes 10 cm in length with a diameter of 0.8 to 1.2 mm. Cleanse the finger and puncture deeply. Discard the first 2 drops of blood. Secure the third drop of blood, note the time and fill the tubes with the blood by capillary attraction. At the end of 3 minutes, break about 1 cm. Length of tubing every 30 seconds and record the clotting time as the interval from the time the blood appeared on the skin until the fibrin thread bridges the broken ends when they have been separated by distance of 5 mm or more. Note the time. The second and third tubes may be used to check the result observed with the first tube. The time between appearance of the blood and filling the tubes and the appearance of the fibrin in coagulation time.

Normal: Varies from 3 to 7 min.

12.11 ERYTHROCYTE SEDIMENTATION RATE (ESR)

Westergren's method

Take 0.4 ml of 3.8% sodium citrate in a tube. Withdraw 2 ml of venous blood in a dry sterile syringe and place exactly 1.6 ml of blood in the tube containing sodium citrate solution. Invert the tube 2 to 3 times to mix the blood with the citrate solution. Fill the Westergren's ESR tube to exactly the 0 mark and place it in the stand. The tube must be held firmly at an exactly vertical position. Reading may be made at 5 minutes at the end of half-an-hour and another at the end of one hour and two hours.

Wintrobe's method

Withdraw 2 ml of venous blood with a dry sterile syringe and place it in the Wintrobe's bottle. Mix gently but thoroughly, with a capillary pipette fill a Wintorbe haematocrit tube to the 10 cm mark with the blood. Keep the filled tube in an exactly vertical position at room temperature between 22°C to 27°C. At the end of one hour read off the height of the clear plasma column above the top of the sinking red cells.

12.12 HAEMATOCRIT VALUE (PACKED CELL VOLUME) (PCV)

Take oxalated blood and mix thoroughly by repeated inversion and fill in Wintrobe's tube up to 100 marks. Centrifuge at 2500 rpm of 30 min, the original column of blood in the tube being 100 mm. The volume of packed cells can be read directly as a percentage.

Normal: Men: 40 – 45%

Women: 36 – 47%

12.13 MEAN CORPUSCULAR VOLUME (MCV)

To determine the average volume of a single and red cell in cubic micron.

$$\frac{\text{Packed cell volume} \times 10}{\text{Red Blood cells in millions per cu. Mm}} = \text{MCV}$$

Normal: 78 to 94 cu.mm. A figure less than 78 is microcytic anaemia. More than 94 is macrocytic anaemia.

12.14 MEAN CORPUSCULAR HAEMOGLOBIN (MCH)

To determine average haemoglobin content of a single red cell in micro-organism.

$$\frac{\text{Haemoglobin} \times 10}{\text{RBC's in millions per cu. mm}} = \text{MCH}$$

Normal: 27 – 32 μg. Less than 27 indicate hypochromic anaemia.

12.15 MEAN CORPUSCULAR HAEMOGLOBIN CONCENTRATION (MCHC)

To determine the haemoglobin content of 100 ml of packed cells as a percentage as opposed to the percentage of haemoglobin of whole blood.

$$\frac{\text{Haemoglobin} \times 100}{\text{Packed cell volume}} = \text{MCHC\%}$$

12.16 FRAGILITY OF ERYTHROCYTES

It is to determine the resistance of the red cells haemolysis in decreasing strength of hypotonic saline.

Prepare 1% sodium chloride aqueous solution and confirm the content by titration and do the correction if necessary. Arrange a series of 12 small test tubes in a rack and number each of them.

Tube no	*1*	*2*	*3*	*4*	*5*	*6*	*7*	*8*	*9*	*10*	*11*	*12*
NaCl 1% in ml	2.0	2.4	2.8	3.2	3.6	4.0	4.4	4.8	5.2	5.6	6.0	6.8
Bring the volume to 8ml by adding distilled water in ml	6.0	5.6	5.2	4.8	4.4	4.0	3.6	3.2	2.8	2.4	2.0	1.2
Concentration of NaCl %	0.25	0.30	0.35	0.40	0.45	0.50	0.55	0.60	0.65	0.70	0.75	0.80
To each tube add 0.2 ml of red cells	←——— 0.2 ———→											
	Mix well											

It is kept at the room temperature for 30 minutes. Then the tubes are centrifuged gently. The degree of haemolysis determined by the supernatant fluid against a standard solution of 100% haemolysis prepared by adding 0.02 ml of oxygenated red cells to 8 ml of ammonia water (0.04%). Similarly a control must be put up side by side with normal blood. The result observed with the control blood should always be reported at the same time.

The normal range of haemolysis will show between:

Tube No. 1 and 2	–	90 to 100%
Tube No. 3, 4 and 5	–	50 to 90 %
Tube No. 6	–	5 to 30%

12.17 FIBRINOGEN

Fibrinogen in plasma was estimated by the method of Ratnoff and Menzie (1951) using the Biuret colour reaction.

Reagents

1. Sodium citrate – 3.18%
2. Calcium chloride – 0.025M
3. Saline – 0.89%
4. **Stock Biuret reagent:** 45g of Rochelles salt was dissolved in about 400ml of 0.2N NaOH to which was added 15g of copper sulphate stirring continuously until the solution is completely dissolved. To the solution, 5g of potassium iodide was added and made upto a litre with 0.2 N NaOH.
5. **Working Biuret solution:** 200 ml of the stock reagent was made upto 1 litre with 0.2 N NaOH containing 5g KI/l.
6. Standard BSA – 6 and aqueous solution of BSA was used as the standard.

Procedure

Two ml of blood was collected into a graduated tube containing 0.4 ml of 3.18% sodium citrate shaken and centrifuged to separate plasma. To 0.5 ml of plasma, 4ml of distilled water and 3 ml $CaCl_2$ were added, mixed and the tube incubated at 37°C for 15 min the fibrin clot shrunk and it was held at the bottom of the tube. The supernatant was discarded and the fibrin clot was washed thrice with 5 ml volumes of saline. The fibrin clot in the bottom was dissolved in 5 ml of working Biuret reagent by warming in a boiling water-bath and 3 ml of saline was added to the tube and optical density measured at 540 nm. Graded volumes of BSA standard ranging from 60 – 240 mg were taken. The final volume was made upto 3 ml with saline. To these tubes 5 ml of Biuret reagent was added and read at 540 nm.

Fibrinogen values were expressed as mg/dl plasma.

Reference

Ratnoff, O.D. and Menzie, C. (1951), J Lab Clin Med., 37: 316.

12.18 PROTHROMBIN TIME

Plasma prothrombin time is done by Quick's one stage method. A potent preparation of rabbit brain emulsion impregnated with calcium is added to citrated plasma and the clotting time estimated.

Reagents

1. **Thromoplastin with calcium:** This is prepared from a commercially available lyophilized rabbit brain preparation.
2. Trisodium citrate – 3.18%

Procedure

In a graduated tube containing 0.4 ml of 3.18% trisodium citrate 1.6 ml of blood was collected shaken and centrifuged to separate the plasma. The plasma is kept at room temperature until the time of the test. The test is performed within 4 hrs of blood collection 0.1 ml of plasma is delivered into the bottom of a short 5 ml test tube (75 mm × 10 mm) and it is placed in the water-bath at 37°C, to this tube 0.2 ml of thromboplastin is added by forcibly blowing from a pipette as quick as possible and a stop watch started simultaneously. The tube is held with its lower end submerged in water (37°C) and gently tilted back and forth until the first strand of fibrin appears. Tilting is done by gently inclining the tube from the vertical just short of horizontal so that the fluid content could be observed. The stop watch is stopped at the time of appearance of the clot and the time noted. This test is repeated once for each specimen and the mean time is recorded. A normal (control) plasma is included with every batch of test samples. Prothrombin time is expressed in seconds.

Reference

Dacie, J.V. and Lewis, S.M. (1972), Ser Haematol. 5: 3.

12.19 BLOOD GROUPING AND CROSS MATCHING – DIRECT AGGLUTINATION

The plasma of a patient may agglutinate or haemolyse the corpuscles of donor or the corpuscles of patient may be agglutinated or haemolysed by plasma of a donor.

Therefore, for the purpose of blood transfusion it is important to do a grouping and cross matching to confirm that no agglutination has taken place between the donor's blood and recipient's blood. There are four major groups:

Blood group A (Anti – B agglutinin in Serum or Plasma)

Blood group B (Anti – A agglutinin in Serum or Plasma)

Blood group AB (Anti – A and B agglutinin in Serum or Plasma)

Blood group O (no agglutinin in Serum or Plasma)

Procedure

Puncture the finger and collect two drops of blood in a small test tube containing about 2 ml of normal saline. Mix gently. Take a glass slide and divide into two with the help of a wax pencil and mark 'A' and 'B'. Put a drop of suspension of blood and a drop of Group 'A' sera in one side of the blood and in the other side a drop of suspension of blood and a drop of Group 'B'sera. Mix gently, by rocking the slide. Allow to stand for 5 minutes, occasionally rolling and tilting the slide to ensure thorough mixing. The cells will then nearly always settle out conspicuously and completely clumped. When no agglutination occurs the cells will remain evenly distributed for hours.

Results

Interpretation

Agglutination of erythrocytes indicated a positive result

Antigen carried by erythrocyte	*Circulating anti-bodies in blood*	*Reaction with anti-A antibody*	*Reaction with anti-B antibody*	*Type of blood group antigen*
'A'	Anti-B	Positive	Negative	'A'
'B'	Anti-A	Negative	Positive	'B'
'AB'	None	Positive	Positive	'AB"
None	Anti – A and Anti – B	Negative	Negative	'O'

Note: The Rh factor of the blood can be determined by using antigen – D.

Reference

Mueller-Eckrardt, C. Blood Group Serology: Laboratory Notes for Medical Diagnostics. Behringwerke A.G. Medizinische, Information/Eukerdruck K.G. Marburg/Lahn, Germany (1975).

Cross Matching

Take one clean slide and divide into two with a wax pencil. Mark 1 and 2.

1. Patients cell suspension 1 drop, donor's serum one drop and the drop of normal saline.

2. Donor's corpuscles suspension, patient's serum and one drop of normal saline.

   Rock the slide occasionally for 5 min and examine under microscope for any agglutination, but the final reading should be made after 15 min. Put up a control side by side of patient's cell and a drop of normal saline in same way the donor's cell with a drop of normal saline.

12.20 IMMUNOGLOBULINS

Quantitation of IgA, IgG and IgM in serum was attained by single radial immunodiffusion techniques by method of Mancini et al (1965).

Reagents

1. Agarose 3% in PBS, pH 7.4.
2. Antisera for IgG, IgA and IgM obtained from Immuno-diagnostic Ltd.
3. Antisera – agarose mixture. Equal volumes of antiserum after required dilution and 3% molten agrose mixed so as to get a final concentration of 1.5% agarose containing antiserum.
4. Phosphate buffer saline, pH 7.4.

Procedure

The antibody-agarose mixture was used to obtain plates of 74 × 74 and allowed to set at 4°C in the cold room for 2 – 3 hours. Then the gel plate was placed on top of a template and wells of 1.5 mm diameter were puched at regular intervals at 1.5 cm. For IgG estimation serum was diluted 5 times with PBS, for IgA, 2 times and for IgM undiluted plasma was used. 1.0 μl of this antigenic solution was delivered into the well with the help of a microsyringe. Graded amounts of standard immunoglobulin solution diluted in PBS, was located in some of the wells. After filling the wells, the plates were kept in a humid chamber at a perfectly horizontal position at room temperature for 3 days. After equilibrium has been attained as a result of Ag-Ab reactions, the diameter (d) of the precipitation ring was measured by means of an immunomeasure scale. A standard curve was contructed bny plotting the diameter (d, in cm) of precipitation rings against the log conc of the standards. The concentrations of the unknown sample were read from the curve.

Immunoglobulin values were expressed as mg/dl serum.

Reference

Mancini G., Carbonara A.O. and Heremans, J.F. (1965), Immunochemistry, 2: 235.

Chapter 13

Biochemical Preparations

13.1 STARCH

Peel potatoes and mince well with distilled water and grind the pulp thoroughly. Filter through a cheese cloth to remove coarse particles. The filtrate thus obtained is opalescent and on standing, starch settles to the bottom. Decant carefully and wash repeatedly. Drain thoroughly and air dry. The same procedure can be employed for isolation of starch from any tuber.

13.2 ESTIMATION OF PURITY OF STARCH

Starch when hydrolysed with acid yields glucose which can be estimated by Hanes method. In this method ferricyanide is reduced to ferrocyanide and unreacted ferricyanide can be found out by iodometric method. From this the amount of ferricyanide reacting with glucose can be calculated thereby the purity of starch can be ascertained.

Reagents

1N hydrochloric acid

20% sodium hydroxide

N potassium dichromate

10% potassium iodide

250 mg of starch

Hanes solution:

Solution A: Dissolve 8.25 g potassium ferricyanide and 10.6 g of anhydrous sodium carbonate in water and make the volume to 100 ml.

Solution B: Dissolve 2.5 g of potassium iodide, 50 g zinc sulphate and 25 g sodium hydroxide and make up the volume to 100 ml with distilled water.

Solution C: 5% acetic acid.

Hydrolysis of Starch

Boil 250 mg of the extracted starch with 50 ml of water and 2 ml of 1N HCl for about 2 h in a boiling water bath. Test the solution for the presence of unhydrolysed starch with iodine solution. Cool the solution, neutralize with sodium hydroxide, using bromothymol blue as an external indicator and make up the volume to 100 ml.

Standardization of Thiosulphate

Titrate 10 ml of standard potassium dichromate solution against thiosulphate solution after adding 5 ml of 10% potassium iodide and 2 ml of conc HCl using starch as indicator. The end point is the appearance of pale green colour.

Purity of Starch

Take 1 ml of hydrolyzed starch solution in a test tube and take 1 ml of water in another and the tubes are kept in a water bath for 20 min. Cool the tubes and transfer the contents into different conical flasks. To the contents of both the flasks add 5 ml of solution A, 5 ml of solution B and 3 ml of solution C. A precipitate is formed on addition of these reagents. Dissolve the precipitate by adding 14 ml of sulphuric acid and titrate the solutions against thiosulphate using starch as indicator. From the titre value the amount of glucose present in the hydrolyzed starch is calculated by multiplying by a factor of 0.93.

Calculation

Volume of thiosulphate consumed by 1 ml of starch

$$(\text{Test titre value} - \text{Blank}) = t \text{ ml}$$

Amount of glucose present in 100 ml of hydrolyzed starch $= t \times 100 \times 0.93$ $= y$

$$\text{Purity of starch} = \frac{y \times 100}{250} = a\% \text{ of starch}$$

13.3 GLYCOGEN FROM RAT LIVER

Sacrifice a well-fed rat and excise the liver immediately. Instant killing is necessary since any form of stress may cause hydrolysis of glycogen. Wash the excised liver in cold saline to remove any blood, quickly blot between folds of filter paper and weigh wet. Keep the tissue in beaker kept in an ice-bath. Now mince and homogenize the tissue after adding 5% TCA solution (2-3 ml per g).

Centrifuge the homogenate at 3000 g for 10 min in cold. Remove the supernatant and repeat the homogenization with the tissue sediment with half the volume of 5% TCA and centrifuge in a refrigerated centrifuge. Pool the supernatant, add twice the volume of 45% ethanol, and after mixing well leave overnight in a refrigerator. This helps to precipitate the glycogen. Centrifuge and collect the precipitate. Dissolve it in a minimal volume of water and reprecipitate as before by adding twice the volume of ethanol. Wash the precipitate with ethanol once and then with ethyl ether, before storing.

13.4 ISOLATION OF LACTOSE FROM MILK

Lactose, milk sugar is isolated from milk by acid and salt precipitation of milk proteins, followed by ethanol treatment.

Procedure

Take 100 ml of milk sample in a 250 ml beaker and add 10% acetic acid solution drop wise until there is significant precipitation of casein. Stir the contents and remove the precipitate by filtration or centrifugation. Collect the filtrate or the supernatant.

To the clear filtrate or the supernatant, add 2.5 – 3.0 g of solid calcium carbonate. Heat the contents by gentle boiling for 10 min over a hot plate. This step will lead to the precipitation of lactalbumin. Cool to room temperature and filter the solution to remove the precipitated albumin and undissolved salt.

Concentrate the filtrate to 10 – 15 ml by gentle boiling. To the hot concentrate, add 85 ml of 95% ethanol and one gram of charcoal. Mix the contents thoroughly and filter. Collect the clear filtrate in a pear shaped flask. Allow the contents to stand for 3 – 5 days at room temperature, so as to allow the lactose to crystallize. Later, collect the lactose crystals by filtration. Wash the crystals with 5 – 10 ml of 25% ethanol. Dry the sugar crystals and record their weight.

13.5 ISOLATION OF GLUTAMIC ACID FROM GLUTEN

Gluten, a major wheat protein is a rich source of glutamic acid. Glutamic acid is isolated from gluten by acid hydrolysis followed by crystallization.

Reagents

Sodium hydroxide, 4% w/v

Congo red indicator, 0.1% w/v in ethanol

Acetone

Hydrochloric acid

Diethyl ether

Procedure

Weigh 50 g of wheat flour and prepare dough with addition of distilled water. Allow it to stand for 30 – 40 min in a closed container. Later, place the dough in a muslin cloth and repeatedly dip and wash with distilled water to remove the starch. Squeeze out the residual water from the gluten and air-dry the sample. Transfer the sample into a 250 ml round bottomed flask and add 30 ml of conc. HCl. Heat the contents on the steam bath until the protein dissolves. Later, reflux the contents for 8 hours using a heating mantle and reflux condenser. To avoid bumping, add pumice stones or porcelain chips. Cool the contents to room temperature and dilute with equal volumes of distilled water.

Concentrate the diluted protein hydrolysate to 10-15 ml by evaporation or over a steam bath using a China dish. Transfer the concentrate to a 150 ml conical flask and saturate the contents by bubbling with dry HCl gas. Cool and seal the contents and leave it in a refrigerator at 4°C for 2 – 3 days, so as to facilitate the process of crystallization of glutamic acid hydrochloride.

Filter the contents after addition of an equal volume of cold acetone using a sintered glass funnel. Wash the crystals with cold acetone followed by diethyl ether. Redissolve the dried sample in minimum volume of hot distilled water and neutralize the solution by adding 4% NaOH drop wise, till it tests positive (use a small aliquot ~ 50 µl for the spot test on a white porcelain tile) with Congo red indicator to give a blue colour (pH 3.0 – 3.3). Centrifuge the sample at 3,000 rpm for 10 min to remove any undissolved impurities. Concentrate the supernatant by evaporation to 5 ml. Allow the glutamic acid to crystallize by placing the sample in a refrigerator for 24-48 h. Recover the crystals by filtration and air dry the sample. Record the weight and calculate the percent yield.

13.6 PREPARATION OF CYSTINE FROM SHEEP'S WOOL OR HUMAN HAIR

The protein keratin is a rich source of sulphur containing amino acids, such as cystine. It is obtained from keratin by acid hydrolysis followed by precipitation with sodium acetate solution at pH 5.0.

Reagents

Formic acid, 50% v/v

Sodium acetate, 50% w/v

Concentrated HCl, 35 - 36%

Dilute HCl, 3%

Chilled diethyl ether

Activated charcoal

Indicator - Congo red

Ammonium hydroxide

Ethanol

Procedure

Weigh 200 g of sheep wool or human hair and wash repeatedly with chilled diethyl ether in a glass tray. Cut the dried specimen into small fragments with a sharp scissors and transfer into a 2 litre round bottomed flask. Add 50 ml of extraction reagent (mix 100 ml of conc. HCl with 400 ml of 50% formic acid) along with a few porcelain chips and reflux the contents over a heating mantle for 18 h.

Transfer the hot contents immediately into a 1 litre glass beaker and add 5 g of activated charcoal. Stir the contents vigorously with a glass rod and filter using filtration unit. Collect the filtrate and concentrate it into thick syrup by evaporation. Dissolve the syrup in 100 ml of distilled water and add 50% hot sodium acetate solution, until the pH reaches 5.0, testing with Congo red for a change in colour from blue to red.

Store the contents at ambient temperature for 72 h in order to precipitate cystine. Recover the material by filtration and wash it with warm distilled water. Dissolve the residue in 200 – 250 ml warm 3% HCl. Mix the contents and filter. Collect the filtrate and add 2 g of activated charcoal. Stir and filter. If any colour persists, repeat the charcoal treatment.

Cystine is precipitated by adding 50% sodium acetate solution, until the pH reaches 5.0. Allow the contents to stand at ambient temperature for 6 h. The precipitated amino acid is filtered and washed with warm water. Transfer the residue on a watch glass/petri plate and dry overnight at 70°C in a hot air over.

Dissolve the dried material in a minimum volume of ammonium hydroxide solution. Remove any undissolved matter by filtration. Cystine present in the clear filtrate is precipitated by cautious addition of conc. HCl. Recover the crystalline precipitate by filtration followed by washing with ethanol and ether. Dry the recovered sample at ambient temperature in a petri plate and weigh the content.

13.7 PREPARATION OF EGG ALBUMIN

Albumin a water soluble protein is isolated from egg white by salting out technique. Globulins precipitate at half saturation, while albumins precipitate at full saturation with ammonium salts.

Procedure

Collect egg white from two eggs carefully, avoiding the egg yolk into a 500 ml beaker. Dilute the egg white to 100 ml by adding distilled water with vigorously beating and stirring. Any precipitate formed at this stage is removed by centrifugation. Add 32.5 g of solid ammonium sulphate in small proportions, at room temperature with gentle stirring. Equilibrate the contents for 15 min at room temperature. The precipitate formed is that of globulins, which is removed by filtration or centrifugation.

Saturate the filtrate or the supernatant by adding solid ammomium sulphate (35.5 g). Leave the contents at room temperature for 30 min with intermittent stirring. The albumin precipitate formed is recovered by centrifugation. Dissolve the precipitate in minimum volume of distilled water and dialyse the protein solution extensively in cold, against distilled water in a 2 litre plastic beaker (two to three changes of distilled water over a period 36 – 48h) so as to remove the salt. Measure the volume of the protein solution after dialysis and add 0.05% (w/v) sodium azide as preservative. Store the protein solution in cold.

13.8 CAESIN FROM MILK

Casein is a mixture of phosphoproteins found in milk to the extent of about 3%. It contains all of the common amino acids and is high in the essential ones. The principle used to isolate the caesin is by adjusting the pH of the milk to 4.8, its isoelectric point. Casein is also insoluble in ethanol and this property is used to remove unwanted fat from the preparation.

Take 100 ml of milk in a beaker and warm to 40°C to this add 70% glacial acid drop wise with stirring. At a certain point a mass of casein separates out. Collect this from the beaker with a glass stirring rod. Do not add any more acid than necessary to obtain a good curd. A few minutes must be given between acid additions to allow for stabilization. Press the mass of casein between tissue paper to remove as much water as possible. Place the mass of cascin in a food blender with 500 ml water. Suspend the casein in the water by blending as much as possible. Separate the casein from the water with a filter and wash with isopropanol. Some low molecular weight alcohol may be used if isopropanol is not available. Let the casein powder air dry.

13.9 LECITHIN FROM EGG

Carefully separate the egg white and yolk in two different beakers. Add 50 ml of ethanol and 25 ml of ethyl ether to the yolk and stir vigorously for a few minutes. Allow it to stand for 15 min with occasional stirring. Filter the suspension through a filter paper moistened with alcohol, and collect the filtrate in a dry beaker. Wash the residue in the filter paper with another 15 ml of alcohol-ether mixture (2:1). Evaporate the filtrate to dryness over a steam bath. After cooling dissolve the residue in 10 ml of ethyl ether. Pour the ether solution into 30 ml acetone, slowly and with gentle shaking. A light yellow precipitate is obtained. Filter and without delay dissolve the precipitate collected in the filtrate in 20 ml of ethanol. To this add 20 ml of alcoholic cadmium chloride solution. After about 15 min, the cadmium salts of lecithin precipitate out. Filter and dry the precipitate in air.

13.10 CHOLESTEROL FROM BRAIN

Take about 100 g of brain tissue (sheep) and blend with 400 ml of acetone. Rinse the blender with acetone and stir the combined homogenate for 10 min. Filter the suspension on a Buchner funnel. Blend the residue with a further 200 ml of acetone as before, filter and combine the filtrates, after removing most of the acetone by distillation under reduced pressure, cool the flask under tap water and collect the crude cholesterol on a Buchner funnel.

Dissolve the crude material in minimum volume of hot ethanol and filter while hot using a fluted filter paper. For this, the collecting flask is placed in a boiling water bath with the filter funnel in position. Air dry the cholesterol and record the yield. If necessary the cholesterol may be crystallized from hot ethanol as above. The purity can be checked by estimating cholesterol.

13.11 HEAMOGLOBIN FROM BLOOD

Collect fresh sheep blood from slaughterhouse into a flask containing Alsevier solution. This solution prevents clotting of blood. The composition of the solution is as follows: 2.05 g of dextrose, 0.89 g sodium citrate, 0.42 g of sodium chloride and 0.05 g of citric acid in 100 ml. Defibrinate the blood by shaking the blood with a few pieces of porcelain and filter through a cheese cloth to remove the sediments.

Centrifuge the filtrate at 3000 rpm for 20 min, and discard the supernatant. Gently suspend the sediment in 0.9% NaCl solution and resediment the cells by centrifugation. Slowly add ethyl ether to the sediment with shaking, till a clear solution is obtained due to the lysis of cells. At this stage add 20 ml of distilled water and mix well. Centrifuge at 3000 rpm for 10 min and collect the supernatant in a beaker kept on ice. Discard the sediment made up of cell membranes. To the supernatant add slowly with gentle stirring ice-cold absolute ethanol. The volume of ethanol is calculated so that the final concentration is 30%. Leave the mixture undisturbed overnight at 0°C, when crystals of haemoglobin start appearing. Carefully decant the supernatant, add a small amount of 25% ethanol, gently mix and leave at 0°C for 3 hours. Decant the supernatant. The crystals are dried in a desiccator over conc. H_2SO_4. They are amorphous and pale yellow.

13.12 ISOLATION OF CAFFEINE FROM TEA

Caffeine is a natural product belonging to a group of compounds called alkaloids. It is found in tea (30 to 75 mg per cup), Coca-Cola (46 mg per 12 oz) and coffee (80-125 mg per cup). It stimulates respiration, the heart and the central nervous system, is a smooth muscle relaxant and a diuretic. It is also addictive.

Hot water swells the tea leaves to release caffeine and other compounds such as tannins. Dichloromethane, CH_2Cl_2, is used to selectively extract the caffeine from the water, which retains most of the other organic compounds. After the CH_2Cl_2 solution is separated, it is passed through some solid Na_2SO_4 to remove traces of water. The CH_2Cl_2 is removed by evaporation to give crude caffeine. Finally the caffeine is purified by recrystallization with ethyl acetate.

Procedure

To make a concentrated solution of tea, boil a solution of 0.3 g Na_2CO_3 in 8 ml of deionized water in a 50 ml beaker. Remove the beaker from the hot plate and add a tea bag (which contains about 2 g of tea); virtually all the water will be soaked up by the tea. Use a 30 ml beaker as a plunger to release the tea into the aqueous solution. Allow it to extract the tea for 3 min. Then press the tea bag while pouring the brown liquid into a test tube, releasing as much liquid from the bag as possible. Return the beaker to the hot plate and add 4 ml of water, warm and repeat the release of tea and add this to the centrifuge tube. Repeat the extraction once again with 4 more ml of water.

Cool the centrifuge tube containing the concentrated tea solution for 2-3 min. in ice water, and then add 1 ml of CH_2Cl_2. Stopper the tube with a cork and shake the mixture vigorously for 15 sec.; remove the stopper and centrifuge.

Prepare a pipet with a cotton plug and pour about 1 cm of Na_2SO_4 above the plug (this is a drying agent). Clamp the pipet vertically above a 25 mL (clean and preweighed to the nearest milligram) Erlenmeyer flask. Hold the centrifuge tube close to the top of the pipet and transfer the CH_2Cl_2 layer from the centrifuge tube into the pipet containing Na_2SO_4 so that it drains directly into the beaker. Cool the tea in the centrifuge tube again, add another 1 ml of CH_2Cl_2, re-extract, centrifuge, and transfer the CH_2Cl_2 layer to the beaker via the pipet filter. Repeat with a third ml of CH_2Cl_2.

Insert a thin wooden boiling stick (splint) into the beaker, and warm it on the steam bath to evaporate the CH_2Cl_2. Remove the stick and weigh the pale yellow solid, crude caffeine.

The caffeine is now purified by sublimation. Fit a test tube with a 2-3 mm thick section of a thin walled rubber hose. The rubber band is slid up the tube so that when it is placed inside the 25 ml Erlenmeyer it is suspended 1-2 mm above the bottom of the Erlenmeyer flask. Remove the tube momentarily and place ice water inside it; make sure the outside of the tube is dry and set it back into the Erlenmeyer flask. The assembly is now buried into a sand-bath heated to 120°C (no stirring is necessary). Periodically replace the tube's water with ice water - but do not spill any water into the flask. If sublimation - the formation of white crystals on the outside of the tube - does not begin within 5 minutes, increase the temperature of the sand. When all the pure caffeine has sublimed, remove the assembly from the sand bath.

To transfer the caffeine, set a preweighed (to the nearest mg) bag into a 400 or 600 ml beaker so that it remains open and able to receive the scraped caffeine from the test tube. Remove the test tube and scrape it thoroughly. Use a melting point capillary to scrape just enough - either from the tube or the bag - for a melting point determination. Then reweigh the bag before it is sealed and submitted.

Reference

"Mayo et al." Mayo, D.W., Pike, R.M., Butcher, S.S. and Trumper, P.K. 1991, Microscale Techniques for the Organic Laboratory; Wiley: New York.

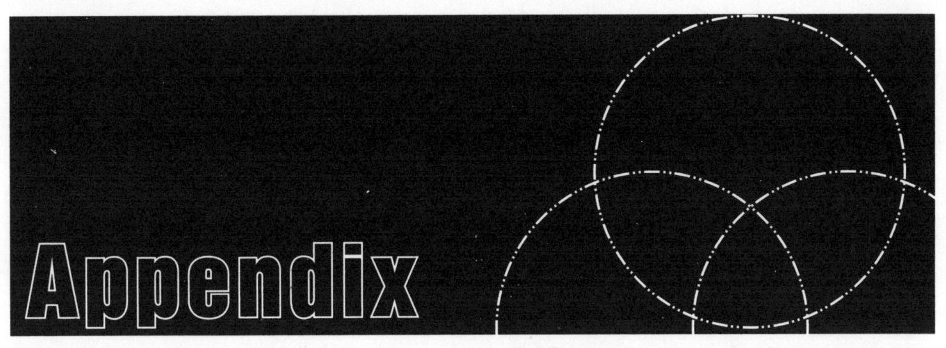

Appendix

BUFFER PREPARATION

1. Glycine – HCl buffer

Stock solutions:

A: 0.2 M solution of glycine (15.01g in 1000ml)

B: 0.2 M HCl

50 ml of A + X ml of B, diluted to total of 200 ml

X	*pH*	*X*	*pH*
5.0	3.6	16.8	2.8
6.4	3.4	24.2	2.6
8.2	3.2	32.4	2.4
11.4	3.0	44.0	2.2

2. Citrate buffer

Stock solutions:

A: 0.1 M solution of citric acid (21.01 g in 1000 ml)

B: 0.1 M solution of sodium citrate (29.41 g $C_6H_5O_7Na_3.2H_2O$ in 1000 ml)

X ml of A = Y ml of B, diluted to a total of 100 ml.

X	Y	pH
46.5	3.5	3.0
43.7	6.3	3.2
40.0	10.0	3.4
37.0	13.0	3.6
35.0	15.0	3.8

33.0	17.0	4.0
31.5	18.5	4.2
28.0	22.0	4.4
25.5	24.5	4.6
23.0	27.0	4.8
20.5	29.5	5.0
18.0	32.0	5.2
16.0	34.0	5.4
13.7	36.3	5.6
11.8	38.2	5.8
9.5	41.5	6.0
7.2	42.8	6.2

3. Acetate buffer

Stock solutions:

A: 0.2 M solution of acetic acid (11.55 ml in 100 ml)

B: 0.2 M solution of sodium acetate (16.4 g of $C_2H_3O_2Na$ or 27.2 g of $C_2H_3O_2Na.3H_2O$ in 1000 ml)

X ml of A + Y ml of B diluted to 100 ml.

X	y	pH
46.3	3.7	3.6
44.0	6.0	3.8
41.0	9.0	4.0
36.8	13.2	4.2
30.5	19.5	4.4
25.5	24.5	4.6
20.0	30.0	4.8
14.8	35.2	5.0
10.5	39.5	5.2
8.8	41.2	5.4
4.8	45.2	5.6

4. Phosphate buffer

Stock solutions:

A: 0.2 M solution of monobasic sodium phosphate (27.8 g in 1000 ml)

B: 0.2 M solution of dibasic sodium phosphate (53.65 g of $Na_2HPO_4.7H_2O$ or 71.7 g of $Na_2HPO_4.12H_2O$ in 1000 ml)

X ml of A + Y ml of B diluted to 200 ml

X	y	pH	x	y	pH
93.5	6.5	5.7	45.0	55.0	6.9
92.0	8.0	5.8	39.0	61.0	7.0
90.0	10.0	5.9	33.3	67.0	7.1
87.7	12.3	6.0	28.0	72.0	7.2
85.0	15.0	6.1	23.0	77.0	7.3
81.5	18.5	6.2	19.0	81.0	7.4
77.5	22.5	6.3	16.0	84.0	7.5
73.5	26.5	6.4	13.0	87.0	7.6
68.5	31.5	6.5	10.5	90.5	7.7
62.5	37.5	6.6	8.5	91.5	7.8
56.5	43.5	6.7	7.0	93.0	7.9
51.0	49.0	6.8	5.3	94.7	8.0

5. Tris (hydroxymethyl) aminomethane (Tris) buffer

Stock solutions:

A: 0.2 M solution of Tris (hydroxymethyl) aminomethane (24.2 in 1000 ml)

B: 0.2 M HCl

50 ml of A + y ml of B diluted to a total of 200 ml

x	pH
5.0	9.0
8.1	8.8
12.2	8.6
16.5	8.4
21.9	8.2
26.8	8.0
32.5	7.8
38.4	7.6
41.4	7.4
44.2	7.2

5. Glycine NaOH buffer

Stock solution:

A: 0.2 M solution of glycine (15.01 g in 1000 ml)

B: 0.2 M NaOH

50 ml of A + y ml of B diluted to a total of 200 ml

x	pH	x	pH
4.0	8.6	22.4	9.6
6.0	8.8	27.2	9.8
8.8	9.0	32.0	10.0
12.0	9.2	38.6	10.4
16.8	9.4	45.5	10.6

6. Carbonate – Bicarbonate buffer

Stock solutions

A: 0.2 M solution of anhydrous sodium carbonate (21.2 g in 1000 ml)

B: 0.2 M solution of sodium bicarbonate (16.8 g in 1000 ml)

50 ml of A + y ml of B diluted to a total of 200 ml.

x	y	pH
4.0	46.0	9.2
7.5	42.5	9.3
9.5	40.5	9.4
13.0	37.0	9.5
16.0	34.0	9.6
19.5	30.5	9.7
22.5	28.0	9.8
25.0	25.0	9.9
27.5	22.5	10.0
30.0	20.0	10.1
33.0	17.0	10.2
35.5	14.5	10.3
38.5	11.5	10.4
40.5	9.5	10.5
42.5	7.5	10.6
45.0	5.0	10.7

7. PM Buffer*

100 ml 1 M PIPES (100 mM PIPES pH = 6.9)

2 ml 1 M $MgSO_4$, (2 mM $MgSO_4$)

2 ml 0.5 M EGTA (1 mM EGTA,)

distilled water to 900 ml
adjust pH to 6.9
distilled water to 1 liter
store at 4°C

8. PBS*

8.18 g NaCl (140 mM NaCl)
0.186 g KCl (2.5 mM KCl)
0.218 g KH_2PO_4 (1.6 mM KH_2PO_4)
2.15 g Na_2HPO_4 (15 mM Na_2HPO_4)
distilled water to 1 liter
Store at room temperature

9. 1 M HEPES, pH = 7.0 Stock*

119.15 g HEPES (free acid)
distilled water to 400 ml
add solid NaOH a few pellets at a time while mixing until the pH is ~6.8
add concentrated NaOH dropwise to achieve pH = 7.0
distilled water to 500 ml
sterile filter and store at 4°C

10. 1 M PIPES, pH = 6.9 Stock*

151.2 g PIPES (free acid)
distilled water to 400 ml
add solid NaOH a few pellets at a time while mixing until the pH is ~6.7
add concentrated NaOH dropwise to achieve pH = 6.9
distilled water to 500 ml
sterile filter and store at 4°C

CHEMICALS

Chemical	*Formula*	*Molecular weight*
Acetic acid glacial	$C_2H_4O_2$	60.05
Acetic anhydride	$C_4H_6O_3$	102.1
Acetone	$C3H_6O$	58.1
Acid hydrochloride	HCl	36.47

Acid nitric	HNO_3	63.0
Acid sulphuric	H_2SO_4	98.08
Alcohol absolute	C_2H_6O	46.07
1-amino-2-mapthol-4-sulphonic acid	$C_{10}H_9O_4SN$	239.24
4-amino antipyrine	$C_{11}H1_3ON_3$	203.25
Ammonium chloride	NH_4Cl	53.49
Ammonium hydroxide	NH_3	17.03
Ammonium molybdate	$(NH_4)MO_7O_{24}.4H_2O$	1235.09
Ammomium phosphate (dibasic)	$(NH_4)HPO_4$	132.1
Ammonium sulphate	$(NH_4)SO_4$	132.14
Barium chloride	$BaCl_2.2H_2O$	244.28
Benzoic acid	C_7H_6O2	61.83
Calcium carbonate	$CaCO_3$	100.1
Calcium chloride	$CaCl_3.6H_2O$	219.1
Chloroform	$CHCl_3$	119.4
Cholesterol	$C_{27}H_{46}O$	386.66
Creatinine	C_4H_7ONa	113.12
Dextrose	$C_6H_{12}O_6$	180.2
2,4-diphenylhydrazine	$(NO_2)C_6H_3.NH.NH_2$	198.14
Disodium phenyl phosphate	$Na_2\ (C_6H5)PO_4.2H_2O$	254.09
Ferric chloride	$FeCl_3.6H_2O$	270.3
Ferrous sulphate	$FeSO_4.7H_2O$	278.0
Formaldehyde	CH_2O	30.0
Glycerin	$C_3H_8O_3$	92.1
Hydrogen peroxide	H_2O_2	34.1
Iodine	I_2	126.9
Magnesium citrate	$Mg_3\ (C_6H_5O_7).H_2O$	469.0
Magnesium acetate	$MgC_4H_6O_4.4H_2O$	214.46
2,4-dinitrophenyl hydrazine	$C_6H_6O_4N_4$	198.14
Nicotinamide adenine dinucleotide (NADH)	$C_{21}H_{27}N_7O_{14}$	709.46
Phenol	C_6H_6O	94.1
Phenolphthalein	$C_{20}H_{14}O_4$	318.3
Phosphoric acid	H_3PO_4	98.05
Picric acid	$C_6H_3O_7N_3$	229.11
Potassium chromate	K_2CrO_4	194.2

Potassium cyanide	KCN	65.1
Potassium dichromate	$K_2Cr_2O_7$	294.2
Potassium dihydrogen phosphate	KH_2PO_4	136.09
Dipotassium hydrogen phosphate	K_2HPO_4	174.18
Potassium ferricyanide	$K_3Fe(CN)_6$	329.25
Potassium ferrocyanide	$K_4Fe(CN)_6.3H_2O$	422.4
Potassium hydroxide	KOH	56.1
Potassium iodide	KI	166.0
Potassium nitrate	KNO_3	101.1
Potassium oxalate	$K_2C_2O_4.H_2O$	184.2
Potassium permanganate	KMnO4	158.0
Potassium sodium tartarate	$KNaC_4H_4O_6.4H_20$	282.2
Potassium thiocyanate	KSCN	97.2
Pyruvic acid	$C_3H_4O_3$	88.0
Sodium acetate	CH3COONa	82.03
Sodium benzoate	$C_7H_5O_2Na$	144.1
Sodium bicarbonate	$NaHCO_3$	84.0
Sodium bisulphate	$NaHSO_4$	120.1
Sodium carbonate	Na_2CO_3	106.0
Sodium chloride	NaCl	58.45
Sodium hydroxide	NaOH	40.0
Sodium nitrite	$NaNO_2$	69
Sodium oxalate	$Na_2C_3O_4$	134.0
Sodium pyruvate	$CH_3COCOONa$	110.05
Sodium sulphate	Na_2SO_4	142.1
Sodium sulphite	Na_2SO_3	126.05
Sodium tungstate	$Na_2WO_4.2H_2O$	329.86
Sulphur	S	32.06
Sucrose	$C_{12}H_{22}O_{11}$	342.3
Tartaric acid	$C_4H_6O_6$	150.1
Trichloroacetic acid	$C_2HO_2C_{l3}$	163.4
Urea	CH_4ON_2	60.1
Uric acid	$C5H_4N_4O_3$	168.11
Zinc sulphate	$ZnSO_4.7H_2O$	287.5

CONCENTRATIONS OF SOME ACIDS AND BASES

Acid or base	*Molecular weight*	*Specific gravity*	*Percent by weights*	*Normality*	*ml/L for 1 N solution*
Acetic acid	60.1	1.05	99.5	17.6	56.9
Ammonium hydroxide	35.0	0.89	28.0	15.1	66.5
Formic acid	46.0	1.20	97.0	25.5	39.2
Hydrochloric acid	36.5	1.18	36.0	11.7	85.5
Nitric acid	63.0	1.42	69.5	15.6	64.0
Perchloric acid	100.5	1.69	70.0	11.6	85.7
Phosphoric acid	98.0	1.69	85.0	44.1	22.7
Sulphuric acid	98.08	1.84	96.0	35.9	28.4
Pyridine	79.1	0.98	100.0	12.4	80.6
2-mercaptoethanol	78.13	1.14	100.0	14.6	68.5

FILTERS FOR COLORIMETRY

Wavelength	Colour of filter used	Colour observed
400	Violet	Greenish yellow
425	Indigo blue	Yellow
450	Blue	Orange
490	Blue green	Red
510	Green	Purple
530	Yellow green	Violet
550	Yellow	Indigo blue
590	Orange	Blue
640	Red	Bluish green
730	Deep red	Green

Determination of Protein Concentration by UV Absorption Method

Pure protein solutions can be quantified by recording the absorbance at 280 nm (A_{280}) or 205 nm (A_{205}). The absorbance of protein at 280 nm is mainly due to tryptophan and tyrosine residues, while absorbance of proteins in the region of 205 – 220 nm is primarily due to the peptide bonds, even though some amino acids may also contribute to this absorbance. This is rapid method of protein quantitation in solution, with no destruction of the sample during analysis. The concentration of the protein (mg/ml) is calculated by the following equation:

$$A205 \div 27 + \left(\frac{A280}{A205}\right)$$

$$[(1.55 \times A280) - (0.76 \times A260)$$

Protein molar conversion

100 pmoles of 100000 MW protein = 10 μg

100 pmoles of 50000 MW protein = 5 μg

100 pmoles of 10000 MW protein = 1 μg

Protein/DNA conversions

1 kb of DNA = 333 amino acids of coding capacity = 3.7 × 104 MW

10000 MW protein = 270 bp DNA

30000 MW protein = 810 bp DNA

50000 MW protein = 1.35 kb DNA

100000 MW protein = 2.7 kb DNA

Determination of Nucleic acid Concentration by UV Absorption

Pure nucleic acid solution shows absorption maxima at 260 nm. This property is typically made use in the absorption in the spectrophotometric determination of nucleic acids. For a nucleic acid solution with an A_{260} = 1, the following approximations are valid,

1 A_{260} unit * of double stranded DNA = 50 μg/ml

1 A_{260} unit * of single stranded DNA = 37 μg/ml

1 A_{260} unit * of single stranded RNA = 40 μg/ml

1 A_{260} unit * of oligonucleotide = 50 μg/ml

*Unit – the concentration of nucleic acid dissolved in 1 ml buffer (20mM sodium phosphate, pH 7.0, 0.1M NaCl), which has an absorbance or 1 (at A_{260}). The spectrophotometric measurement is made in a 1cm cuvette, at 20°C.

Further, the ratio A_{260}/A_{280} provides a good index of the purity of a nucleic acid preparation. A pure DNA and RNA have A_{260}/A_{280} ratio of 1.8 and 2.0 respectively. If the ratio is found to be less than 1.8 or 2.0 then it can be inferred that the preparation is contaminated with protein or phenol.

Troubleshooting in Molecular Biology

SDS PAGE gel electrophoresis

Trouble	*Causes*	*Remedy*
Incomplete polymerization	Presence of oxygen	Degas the solutions sufficiently Seal the top edge to be polymerized with gel buffer or butanol instead of water
	Stock solution old	Always store stock solutions in the refrigerator. Persulphate solution should be prepared fresh.
	Improper preparation of solutions	Check if all solutions are mixed and prepared well.
Irregular shaped or distorted sample wells	Air bubbles trapped under comb	Move the combs until the air bubbles rise up and place the combs back.
	Combs removed before polymerization is complete	Keep little amount of mixture in a beaker to check for complete polymerization.
	Stacking gel and comb	Fit and remove the comb carefully.
Extended running time	Chloride ions in cathode buffer	Do not use HCl to bring the pH of Tris-glycine to pH 8.9
	Concentration of buffer may be high/operating current too low	Adjust the concentration/current
	Air bubbles interference	Flush air bubbles
Reduced running time	Concentration of the buffer may be low/operating current too high	Adjust the concentration/current
Curved dye front ends upwards smiling effect	Uneven electrical conductance usually due to higher temperature at the centre of the gel	Precool the buffer tank in the refrigerator or perform the run in a cold room Reduce the current flow
Ends turned downwards frowning effect	Non uniform electric field	Remove air bubbles from the loser edge of the gel.
Poor resolution of bands	Diffusion before electrophoresis	Start the electrophoresis as soon as sample has been loaded
	Diffusion after electrophoresis	Use higher percentage of acrylamide
		Make sure that gel is completely polymerized and butanol/gel buffer is completely poured off before pouring stacking gel.
Protein bands are wavy	Excess persulphate	Use optimum concentration of persulphate
Protein bands are streaky	Protein remain aggregated denatured or insoluble	Use fresh sample buffer or extra SDS or centrifuge the sample extract
Appearance of double bands	Partial oxidation of proteins	Oxidation can be reduced by increasing 2-mercaptoethanol concentration in the sample buffer.

Uneven protein dye migration	Air bubbles insulation	Remove air bubbles before electrophoresis
	Insuffucient cooling/high current.	
Protein band lane broadens at the bottom of separating gel	Sample density	Load equal volume of samples in each well; equal strength sample buffer; leave no empty wells in the middle
Sample diffuses while loading the wells	Sample density low	Increase sucrose/glycerol in the sample buffer
Poor staining	Dye absorption is not efficient	Dye may be old or change to a more sensitive stain.
Patchy staining	Solid dye	Dissolve or filter the dye
Stained bands are decolourized	Excessive removal of dye	Restain the gel and stop destaining appropriately
Precipitate on the edge of the well	Proteins are too large to pass into the gel pores	Use a gradient gel for separation of larger range of molecular weights.

DNA AGAROSE GEL ELECTROPHORESIS

Trouble	*Causes*	*Remedy*
Faint, smeared or no band on the gel	Insufficient quantity or concentration of DNA loaded on the gel	Increase the amount of DNA, but do not exceed 50 ng/band.
	The DNA was degraded	Avoid nuclease contamination.
	The DNA was electrophoresed off the gel	Decrease the time of the run Use less voltage or higher percent of gel.
	Too much salt in the DNA or protein contamination	Use ethanol precipitation to remove excess salts, before electrophoresis.
	Improper visualization of gel	Use shortwavelength (254 nm).
	Small diffused DNA bands	Add EtBr during electrophoresis.
Anomalies in band migration	Improper electrophoresis condition	Do not allow voltage to exceed ~ 20 V/cm. Maintain temperature < 30°C. Check buffer capacity of the electrophoresis buffer.
	DNA denatured	Do not heat standards prior to electrophoresis. Dilute DNA standards in buffer with 20 mM NaCl.

References

Gabb, M.H. and Latchem, W.E. (1967). *Handbook of Laboratory Solutions.* Andre Deutsch Ltd. London.

Segel, I.H. (1976). *Biochemical Calculations.* 2nd ed., Wiley, New York.

Plummer, D.T. (1971). *An Introduction to Practical Biochemistry,* Tata McGraw-Hill Publishing Company.

Jayaraman, J. (1992). *Laboratory Manual in Biochemistry,* Wiley Eastern Limited, New Delhi.

Wilson, K. and Walker, J. (1996). *Practical Biochemistry-Principles and Techniques,* 4th ed., Cambridge University Press, Cambridge, London.

Parish, J.H. (1972). *Principles and Practice of Experiments with Nucleic acids,* Longman Group Ltd., London.

Augustine, J., Klein, B.P., Becker, D. and Venugopal, P.B. (1985). *Methods of Vitamin Assay,* 4th ed., John Wiley & Sons, New York.

Wood, E.J. (1989). *Practical Biochemistry for Colleges,* Pergamon Press, New York.

Boyer R. (1986). *Modern Experimental Biochemistry,* Addison Wesley Publishing Company.

Hames, B.D. and Rickwood, D. (1990). *Gel Electrophoresis of Proteins – A practical approach,* 2nd ed., IRL Press at Oxford University Press, Oxford, London.

Rickwood, D. and Hanes, B.D. (1990). *Gel Electrophoresis of Nucleic Acids – A practical approach,* 2nd ed., IRL Press at Oxford University Press, Oxford, London.

Sambrook, J., Fritsch, T. and Maniatis, J. (1989). *Molecular Cloning – A laboratory manual,* 2nd ed., Cold Spring Harbour Laboratory Press, London.

Boffey, S. (1996). *Molecular Biology Techniques. In principles and techniques of practical biochemistry.* (Wilson, K. and Walker, J., ed) 4th ed., Cambridge University Press.

Risley, J. (1991). *Preparing Solutions in Biochemistry.* Lab. *J. Chem. Educ.* 68: 1054.

Boyer, R. (1999). *Concepts in Biochemistry.* Brooks/Cole (Pacific Grove, CA),

Rickwood, D. (1993). *Centrifugation – A practical approach,* 3rd ed., IRL Press (Oxford).

Montogomery, R. and Swenson, C.A. (1969). *Quantitative Problems in Biochemical Sciences,* W.H. Freeman & Co. San Francisco, USA.

Dawes, E.A. *Quantitative Problems in Biochemistry.* 6th ed. Longman, London, UK.

Beynon, R.J. and Esterby, J.S. (1996). *Buffer Solutions the Basics.* IRL Press, Oxford.

Vogel, A.I. (1982). *Text of Quantitative Analysis.* 4th ed. ELBS, London, UK.

Day, R.A. Jr. and Underwood, A.L. (1993). *Quantitative Analysis,* 6th ed. Prentice-Hall of India, New Delhi.

Holme, D.J. and Peck, H. (1998). *Analytical Biochemistry.* 3rd ed. Addison Wesley Longman Ltd. Essex, UK.

Varley, H., Gowenlock, A.H. and Bell, M. (1980). *Practical Clinical Biochemistry.* 5th ed. William Heinemann Medical Books Ltd. London.

Carter, H.E. (1950). *Biochemical Preparations.* Vol I. John Wiley & Sons. Inc. NY. USA/Chapman & Hall, London, UK.

Hawk, P.B., Oser, B.L. and Summerson, W.H. (1966). *Practical Physiological Chemistry,* 14th ed. Churchill London, UK.

Bergermeyer, H. U. (ed-in-chief) (1983). *Methods of Enzymatic Analysis.* Vol III & IV. 3rd ed. Verlag Chemie, Weinheim, Germany.

Eisenthal, R. and Danson, N.J. (1992). *Enzyme Assays: A practical approach.* IRL Press, Oxford, UK.

Hawcroft, D. (1987). *Diagnostic Enzymology.* (ed). James, A.M., John Wiley & Sons, Chicherster, UK.

Tietz Fundamentals of Clinical Chemistry. 5th ed. Burtis, C.A., and Ashwood, E.R. (eds). W.B. Saunders Company/Harcourt (India) (P) Ltd., New Delhi, (2001).

Kirk, R.S. and Sawyer, R. (1991). *Pearson's chemical Analysis of Food.* 9th ed. Longman Scientific and Technical; Harlow, Essex, UK.

Miller, D.D. (1998). *Food Chemistry: A laboratory manual.* John Wiley & Sons Inc. NY, USA.

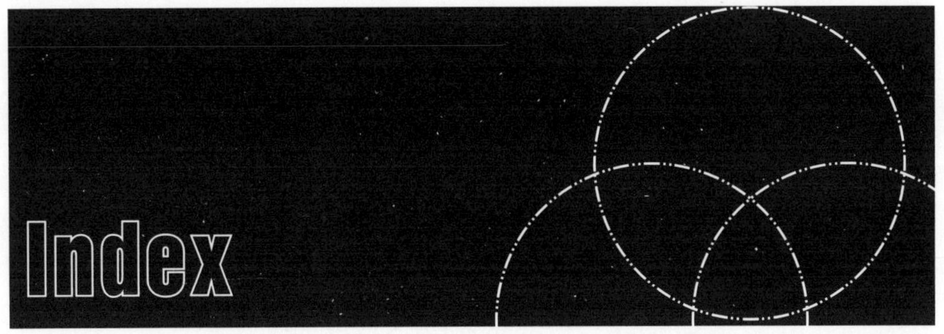

Index